INTERNATIONAL

REVIEW OF CYTOLOGY

VOLUME 94

INTERNATIONAL

Review of Cytology

EDITED BY

G. H. BOURNE
St. George's University School of Medicine
St. George's, Grenada
West Indies

J. F. DANIELLI
(Deceased April 22, 1984)

ASSISTANT EDITOR

K. W. JEON
Department of Zoology
University of Tennessee
Knoxville, Tennessee

VOLUME 94

Part A: Plant
Chromosome Ultrastructure

EDITED BY

G. P. CHAPMAN
Department of Biological Sciences
Wye College
Wye, Ashford, Kent, England

Part B: General Topics

ACADEMIC PRESS, INC. 1985
(Harcourt Brace Jovanovich, Publishers)
Orlando   San Diego   New York   London
Toronto   Montreal   Sydney   Tokyo

ACADEMIC PRESS, INC.
Orlando, Florida 32887

United Kingdom Edition published by
ACADEMIC PRESS INC. (LONDON) LTD.
24–28 Oval Road, London NW1 7DX

LIBRARY OF CONGRESS CATALOG CARD NUMBER: 52-5203

ISBN 0–12–364494–1

PRINTED IN THE UNITED STATES OF AMERICA

85 86 87 88   9 8 7 6 5 4 3 2 1

9/24/85

Contents

Part A. Plant Chromosome Ultrastructure

Plant Chromosomes: A Perspective

G. P. CHAPMAN

The Chromosomes of Dinoflagellates

JOHN D. DODGE

Chromatin Organization and the Control of Gene Activity

WALTER NAGL

Structural and Functional Aspects
of Nucleolar Organizer Regions (NORs)
# of Human Chromosomes

K. A. BABU AND R. S. VERMA

Sertoli Cell Junctions: Morphological
and Functional Correlates

LONNIE D. RUSSELL AND R. N. PETERSON

Functioning and Variation of
Cytoplasmic Genomes: Lessons from
Cytoplasmic-Nuclear Interactions
Affecting Male Fertility in Plants

MAUREEN R. HANSON AND MARY F. CONDE

Ontogeny and Evolution
of Salmonid Hemoglobins

N. P. Wilkins

Contributors

Numbers in parentheses indicate the pages on which the authors' contributions begin.

K. A. BABU (151), *Division of Cytogenetics, Interfaith Medical Center, Brooklyn, New York 11238*

CHING H. CHANG (127), *Department of Cell Biology, Baylor College of Medicine, Houston, Texas 77030*

G. P. CHAPMAN (1, 107), *Department of Biological Sciences, Wye College, Wye, Ashford, Kent TN25 5AH, England*

MARY F. CONDE (213), *Experimental Plant Genetics, The Upjohn Company, Kalamazoo, Michigan 49001*

JOHN D. DODGE (5), *Botany Department, Royal Holloway College, University of London, Egham, Surrey TW20 0EX, England*

M. B. E. GODWARD (77), *School of Biological Sciences, Queen Mary College, University of London, London E1 4NS, England*

GY. HADLACZKY (57), *Institute of Genetics, Biological Research Center, Hungarian Academy of Sciences, H-6701 Szeged, Hungary*

MAUREEN R. HANSON[1] (213), *Department of Biology, University of Virginia, Charlottesville, Virginia 22903*

LARRY I. LIPSHULTZ (127), *Department of Cell Biology, Baylor College of Medicine, Houston, Texas 77030*

LATTA MURTHY (127), *Department of Cell Biology, Baylor College of Medicine, Houston, Texas 77030*

WALTER NAGL (21), *Division of Cell Biology, Department of Biology, The University of Kaiserslautern, D-6750 Kaiserslautern 1, Federal Republic of Germany*

R. N. PETERSON (177), *Department of Physiology, School of Medicine, Southern Illinois University, Carbondale, Illinois 62901*

[1]Present address: Section of Genetics and Development, Cornell University, Ithaca, New York 14853.

DAVID R. ROWLEY (127), *Department of Cell Biology, Baylor College of Medicine, Houston, Texas 77030*

LONNIE D. RUSSELL (177), *Department of Physiology, School of Medicine, Southern Illinois University, Carbondale, Illinois 62901*

DONALD J. TINDALL (127), *Department of Cell Biology, Baylor College of Medicine, Houston, Texas 77030*

R. S. VERMA (151), *Division of Cytogenetics, Interfaith Medical Center, Brooklyn, New York 11238*

N. P. WILKINS (269), *National University of Ireland, University College, Galway, Ireland*

Plant Chromosomes: A Perspective

G. P. Chapman

Department of Biological Sciences, Wye College, Wye, Ashford, Kent, England

At the simplest level of classification chromosomes of organisms with a wall-bound protoplast fall into three groups.

1. Neither nucleosomal nor linear nor membrane limited (bacteria and blue green algae).
2. Neither nucleosomal nor linear but membrane limited (dinoflagellates).
3. Nucleosomal, linear, and membrane limited (eukaryotes).

Beyond this, subdivision into "primitive" and "advanced" is possible and forms the subject of stimulating reviews by, for example, Cavalier-Smith (1981) and Picket-Heaps (1974). For perspective here, however, a probable sequence of the major evolutionary events involving the nucleus could be the following.

1. A prokaryote ancestor had a closed ring of DNA where the terminus of replication (T.O.R.) associated with the cell membrane.
2. Invagination of the cell membrane gave a persistent nuclear envelope with which the T.O.R. continued to associate.
3. Elaboration of the chromosome involved evolution of plural replicons, linearization together with a mechanism for perpetuating "ends" (telomeres) and evolution of nucleosomes.
4. Fenestration of the nuclear envelope gave the kinetochore a role that linked the original (?) T.O.R. with cytoplasmic elements.
5. Diversification of the chromosome by duplication and rearrangement prompted the elaboration of both homology and nonhomology.
6. Elaboration of function gave split genes, noncoding sequences, gene amplification, partition of the chromosome into eu- and heterochromatin, and the emergence of a (probably persistent) nucleolus.
7. The kinetochore became integral to the chromosome and thus a "centromere."
8. Fenestration of the nuclear membrane was taken virtually to total disin-

1

tegration at metaphase and was linked with a sharply defined condensation cycle and a nonpersistent nucleolus.

9. Centromere activity was in some cases dispersed to several parts of the chromosome.

10. Specialization of chromosomes at particular points in the life cycle occurred, examples being emergence of single chromatid types from interphase nuclei for meiotic pairing, multistrandedness in secretory tissues, and loss of totipotency in some aged cells.

Numerous rearrangements of such a sequence are possible but it seems unlikely for example that a nonpersistent nucleolus originally preceded a persistent one or that dispersed centromere activity preceded a nondispersed one however much subsequent evolution reversed these directions. Conversely the elaboration of homology and the evolution of noncoding sequences may or may not be regarded as independent or unrelated events.

The events outlined from 1 to 10 could lead equally to chromosomes of multicellular plants or animals. All higher plant chromosomes however operate within a wall-bound protoplast and additionally, those plants that are green include not only mitochondrial but chloroplast DNA within their cell environment, the inference being that unlike animal chromosomes, those of plants intimately coevolve not with one but two types of subsidiary organelle.

At two extremes detail is readily available. The optical microscopy of plant chromosomes and the base sequences of numerous examples of plant gene are to hand but between these extremes the organization of the chromosome as an arena for molecular events is but little understood and among the questions that could be asked, perhaps the following presently are the most relevant and thus guided the choice of subject matter in the first part of this volume.

1. In the evolution of linear nucleosomal chromosomes do dinoflagellates provide evidence of an intermediate type or are they examples only of an "alternative chromatin"?

2. How are distinctively "plant" characteristics organized and how does plant and animal chromatin compare?

3. How physically manipulable are plant chromosomes and to what extent can techniques for animal chromosomes be applied to them?

4. Lest conclusions be based on too few or unrepresentative samples, how varied is plant chromatin ultrastructurally?

5. How do the centromere and the telomeres, the major "suborganelles" vary?

6. At the applied level what scope is there for *directed* alteration likely to be useful in plant breeding?

As the answers, partial at first, to these questions begin to accumulate, so the study of plant chromosomes can contribute to resolving the enigma of chromosome phylogeny set out earlier.

REFERENCES

Cavalier-Smith, T. (1981). *Soc. Gen. Microbiol. Symp.* **32,** 33–83.
Pickett-Heaps, J. (1974). *Biosystems* **6,** 37–48.

INTERNATIONAL REVIEW OF CYTOLOGY, VOL. 94

The Chromosomes of Dinoflagellates

JOHN D. DODGE

*Botany Department, Royal Holloway College, University of London,
Egham, Surrey, England*

I. Introduction

The chromosomes of dinoflagellates have long been suspected of being unusual. The nucleus is generally big and conspicuous (Fig. 1) and, in the larger species at least, chromosomes can be seen even without any specal staining. The nucleus has been described as a dinocaryon. Early work on nuclear division, such as that of Hall (1925), revealed that the chromosomes exist in a permanently condensed state. In the late 1950s electron microscopy revealed a quite unique ultrastructure to the chromosomes and cytochemical studies showed that their major compenent is DNA. It is now known that no true histones are present and histone-like protein represents only a small proportion of the chromosome. Compared to other organisms dinoflagellates appear to contain a disproportionately large amount of DNA which may be as much as 200 pg per cell in *Gonyaulax polyedra* as compared with only about 5 pg in human cells (Allen *et al.*, 1975). Consequently, dinoflagellate chromosomes have been the subject of many investigations and a fair amount of speculation. In this article the historical work will be reviewed and an attempt made to bring together a current view on the structure of the chromosome.

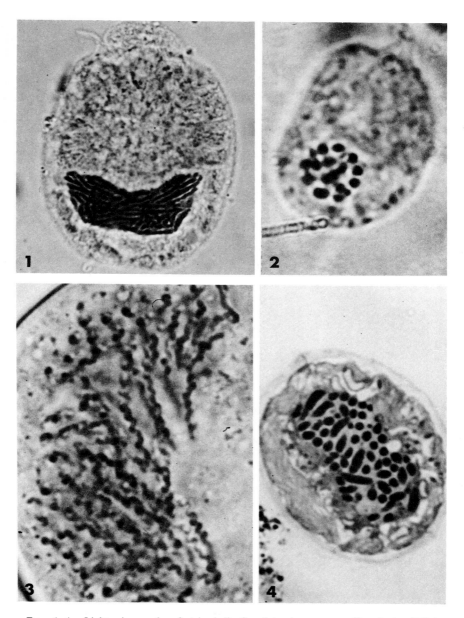

FIGS. 1–4. Light micrographs of stained dinoflagellate chromosomes. Fig. 1. *Amphidinium herdmanae* showing the numerous long chromosomes. Acetocarmine. ×2000. Fig. 2. *A. carterae* with a small number of short chromosomes. Acetocarmine. ×3500. Fig. 3. The long chromosomes of *Prorocentrum micans* which have been treated to reveal evidence of a chromonema which is thinner than the chromosome. Acetocarmine. ×2500. Fig. 4. A section of a plastic-embedded cell of *P. micans* stained with Azur B to show the presence of DNA in the chromosomes. ×2000.

II. Chromosome Numbers

Following the life history studies of von Stosch (1973) it is generally thought that the normal motile dinoflagellate cell has a haploid chromosome number. Nevertheless, the numbers recorded, apart from those of a few parasitic dinoflagellates, are extremely high and because of the difficulty of making squash preparations in general the counts given are only approximate. There is no metaphase plate in the sense that this is found in higher eukaryotes and the counts have generally been made on interphase nuclei where the chromosomes, when long, may be tangled together (Fig. 3). The chromosomes also appear to be rather easily fragmented, perhaps a consequence of the lack of histone in their construction. Recently, special techniques involving cell lysis and enzymic digestion have been devised to assist in spreading and counting these chromosomes (Holt and Pfiester, 1982; Loper *et al.*, 1980) in species where the chromosomes are long and numerous.

The earliest known count (approximation) was that of Borgert (1910) who found that *Ceratium tripos* had about 200 chromosomes. A freshwater species, *C. hirundinella*, was found to have an even higher number, 264–284 (Entz, 1921) but an *Amphidinium* species had only 25 ± 1 (Grassé and Dragesco, 1957). Dodge (1963b) presented interphase counts of acetocarmine-stained nuclei of 11 species. These ranged from 18–22 in *Prorocentrum balticum* to 134–152 in *Gonyaulax tamarensis*. It was of interest that in the genus *Prorocentrum* where five species were counted, there was a clear range of chromosome numbers: *P. balticum*, 20; *P. triestinum*, 24; *P. pusilla*, 24; *P. mariae-lebouriae*, 32; *P. micans*, 68. These could possibly represent different levels of ploidy although the size of the chromosomes was also fairly variable. Among other marine dinoflagellates which have been examined are the much used experimental organism, *Crypthecodinium cohnii*, with 99–100 chromosomes (Allen *et al.*, 1975), *Scripsiella sweeneyae*, with 80–90 (Fine and Loeblich, 1976), and *Heterocapsa pygmaea*, with 61–65 (Loeblich *et al.*, 1981). Chromosome numbers of numerous freshwater dinoflagellates have been counted, particularly for species from India (Sarma and Shyam, 1974; Shyam and Sarma, 1978) and the United States (Holt and Pfiester, 1982). In the case of the Florida red-tide organism, *Gymnodinium breve*, counts of field material and recent isolates were 121 ± 3 but a 25-year-old isolate gave a count of approximately twice as many chromosomes (240 ± 6) and is thought to represent an autodiploid state (Loper *et al.*, 1980).

The shape and size of dinoflagellate chromosomes vary considerably from the small almost oval bodies of *Amphidinium carterae* (Fig. 2) to the long 1-μm-wide threads of *Prorocentrum micans* (Fig. 3). In general, the chromosomes all appear similar within any species and it is arguable that any variations, apart perhaps from those of nucleolar organizing chromosomes, result from the preparation techniques. However, Gavrila (1977) has attempted a karyotype analysis of the 87 chromosomes of *Peridinium balticum* which appear to range in shape from small spheres to long threads.

III. Chemical Composition of the Chromosomes

The use of standard cytochemical tests such as Feulgen and azur B stains (Fig. 4) clearly shows the presence of substantial amounts of DNA in dinoflagellate chromosomes (Ris, 1962; Dodge, 1964). However, attempts to demonstrate proteins were unsuccessful, leading to the conclusion that the chromosomes lack histone. It should be noted, though, that fast green staining has indicated the presence of substantial amounts of basic protein in the parasitic organism *Syndinium* (Ris and Kubai, 1974) and the rather unusual and possibly primitive dinoflagellate *Oxyrrhis* (Hollande, 1974). Many molecular biological techniques have now been used in investigation of the basic components of the chromosomes.

A. DNA

One of the first discoveries concerning the DNA of dinoflagellate chromosomes was that it has an unusual composition. In analyzing the DNA from *Crypthecodinium cohnii* Rae (1973) found a discrepancy in both the buoyant density in CsCl and in thermal gradient studies. He then discovered that this was caused by the replacement of some of the usual base thymidine by 5-hydroxymethyluracil which is not known from any other eukaryotes. This was subsequently confirmed for a number of other dinoflagellates (Rae, 1976). Recently, it has been shown (Herzog and Soyer, 1982a) for *Prorocentrum micans* that 5-hydroxymethyluridylate represents 13.4% of the total nucleotides replacing 63% of the expected thymidine.

Studies of the renaturation kinetics of dinoflagellate DNA (Allen *et al.*, 1975; Roberts *et al.*, 1974; Loeblich, 1976) have shown that there is a clear difference to the DNA of bacteria. It is also suggested that, on the basis of the curve obtained, there is no evidence that the chromosomes are highly polytene for the portion of repeated DNA (of 500–600 base pairs) is interspersed with unique DNA representing approximately 40% of the whole, as happens in higher eukaryotes. The question of polyteny has been disputed by other workers who suggest that in *Prorocentrum micans* the degree of polyteny is about 700 (Haapala and Soyer, 1974). More recent studies of the DNA sequence in *Crypthecodinium* have shown that roughly half the genome is made up of unique sequences which are interspersed with repeated sequences representing about

600 nucleotides (Hinnebusch *et al.*, 1980). Whereas in most respects the arrangement is as in higher plants and animals an unusual class of heteroduplexes was detected by electron microscopy. These are thought to represent the reassociation of repeated sequences from different families and may indicate that there is an unusual organization within the dinoflagellate chromosome.

B. PROTEIN

The first study on isolated dinoflagellate chromatin (Rizzo and Nooden, 1972) showed that although histones were not major constituents, there was a small amount of acid-insoluble protein. Subsequent studies have shown that this protein gives an acrylamide gel banding pattern quite different from that of typical histones (Rizzo and Nooden, 1974) and it has a molecular weight of about 16,000. It is a basic protein differing from histone in containing both cysteine and aromatic amino acids.

C. METALS

Studies on dinoflagellate chromosomes using X-ray microanalysis have resulted in the somewhat surprising discovery of the presence of fairly high levels of transition metals: iron, nickel, copper, and zinc together with chromatin-associated calcium (Kearns and Sigee, 1980; Sigee and Kearns, 1981b, 1982). When similar analyses were carried out on *Glenodinium foliaceum*, a dinoflagellate which contains a eukaryotic endosymbiont, it was discovered that whereas the dinoflagellate nucleus contained the transition elements the nucleus of the symbiont lacked iron and nickel (Sigee and Kearns, 1981a,c).

In an effort to discover the function of the transition metals autoradiographic experiments have been carried out on the uptake of nickel into the chromosomes (Sigee, 1982). Cells labeled for 2 hours showed active uptake throughout the population and 83% of the label was over the dinoflagellate nucleus. The function of transition metals in these particular nuclei is now thought to be related to the stabilization of the chromosome structure in which they may act to form ionic bridges between nucleic acid and the protein matrix (Kearns and Sigee, 1980). It may also be that they form an important structural linking agent within the chromatin which is necessary because of the absence of histones (Sigee, 1983). This suggestion is perhaps at variance with a recent study (Herzog and Soyer, 1983) which suggests that divalent cations (Ca^{2+}, Mg^{2+}) are mainly responsible for the stabilization of the permanently condensed dinoflagellate chromosome architecture. When isolated chromosomes were incubated in EDTA the structure of the chromosomes collapsed and the fibrils (nucleofilaments) separated.

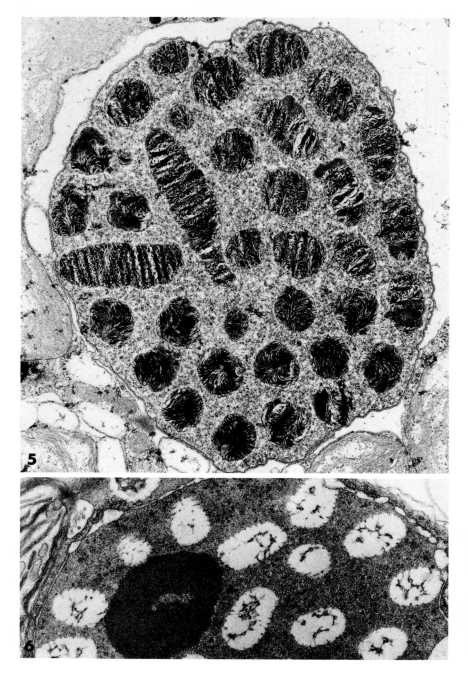

IV. Structural Organization

A. CHROMATIN

The early electron micrographs (e.g., Grell and Wohlfarth-Bottermann, 1957) revealed a fibrillar substructure to the dinoflagellate chromosome which was quite distinct from anything that had hitherto been seen in eukaryotes but was highly reminiscent of the appearance of micrographs of bacterial nuclei (e.g., Ris, 1962). Numerous studies of sectioned chromosomes have subsequently shown the reality of the fibrillar substructure, although the thickness and density of the fibrils vary considerably according to the fixation techniques. Fairly early on (Leadbeater, 1967) it was shown that the interpolation of deoxyribosenuclease extraction following prefixation completely removed the fibrils (Fig. 6) leaving only the spaces formerly occupied by the chromosomes. This test has been repeated by other workers in different dinoflagellates using variations on the technique but with the same result (e.g., Sigee and Kearns, 1981c) that the fibrils appear to consist of DNA. More recently, the chromatin has been studied by spreading techniques. These were pioneered by Haapala and Soyer (1973) and have subsequently been utilized by Hamkala and Rattner (1977), Oakley and Dodge (1979), Rizzo and Burghardt (1980), and Herzog and Soyer (1980). All studies have shown that unlike the normal eukaryotic chromatin, which is beaded in appearance due to the presence of nucleosomes, dinoflagellate chromatin forms a smooth thread with a diameter of about 6.1 nm. Again, comparisons between the two nuclei of a dinoflagellate with an endosymbiont reveal chromatin with nucleosomes from the symbiotic nucleus and smooth fibrils from the dinoflagellate nucleus (Rizzo and Burghardt, 1980; 1982). In the most recent study of dinoflagellate chromatin (Herzog and Soyer, 1980) the appearance of the fibrils has been found to be consistent although a range of isolation techniques were utilized.

B. THREE-DIMENSIONAL STRUCTURE

From the earliest electron microscopic studies it has been evident that dinoflagellate chromosomes are constructed in a precise but distinctive manner. Unfortunately, because of the limitations of fixation and sectioning techniques and the small size ($\sim$ 6nm) of the basic fibrils it has proved extremely difficult to

---

FIG. 5. A thin section through the nucleus of *Amphidinium herdmanae* (cf. Fig. 1) showing chromosomes cut longitudinally and transversely. Conventional TEM fixation, lead and uranyl stain. ×17,500.

FIG. 6. Part of a nucleus of *Glenodinium hallii* which has been treated with DNase during preparation. The chromosomes are now represented by cavities in the nuclear matrix. ×24,000.

interpret the structure of the chromosomes. Nevertheless, there have been many attempts and in the historical account which follows the main models will be briefly described.

The first proposal (Grassé and Dragesco, 1957) was that the chromosome consists of a spiral chromonema made up of numerous fibrils. Next, Giesbrecht (1961) suggested that there was a supercoiled system but he later (1965) modified this to a serpentine coiled system in which the loops of chromonema double back and are held in place by a longitudinal structure at the side of the chromosome. Further studies by Grassé et al. (1965a,b) led to a revision of their earlier model. This really was a hybrid between the earlier spiral model and the serpentine model and tried to take account of the supposed pathway of the fibrils (which they thought represent chromonemas) as seen in thin sections. In 1968 Bouligand et al., using essentially the same type of information on the appearance of the chromosomes in thin section, produced a radically different interpretation based on a mathematical analysis of the arrangement of fibrils. In simple terms this model proposed that the chromosome essentially consists of a number of discs in which the fibrils lie at right angles to the long axis of the chromosome. It was thought that there could be continuity of fibrils between the discs but the overall arc-like arrangement of the fibrils was thought to be an illusion.

The development of chromosome spreading techniques in the early 1970s made possible a fresh viewpoint on the structure of the chromosome. In the first of these studies (Haapala and Soyer, 1973, 1974) it became clear that the chromosome consists of a circular chromatid (or chromatids) twisted together in a spiral in *Prorocentrum micans* or in a regular folding pattern in *Crypthecodinium cohnii*. What was not in dispute was the fact that the ends of the chromosomes are rounded suggesting that the DNA forms continuous loops. This point was further emphasized in studies, by spreading and serial thin sectioning, of the small chromosomes of *Amphidinium carterae* (Oakley and Dodge, 1979). These showed very clearly that the toroidal bundle of chromosome fibrils is plectonemically coiled into a tight double helix (Fig. 7). A different slant on the interpretation of chromosome structure has been provided by Livolant and Bouligand (1978) who, referring back to the earlier work of Bouligand et al. (1968), have likened the chromosome to a liquid crystal of the cholesteric type. All other interpretations are dismissed as being based on artifacts or misinterpretations. In subsequent papers (Livolant, 1978; Livolant and Bouligand, 1980) these authors suggest that it is the effect of the presence of water which causes the chromosomes to extend and give rise to the double helix of the chromonema and the appearance of DNA filaments. This interpretation does seem to be entirely out of line with the interpretations of DNA structures from other organisms and indeed, if it is at all correct, then will cast doubt on all work using spread DNA techniques.

Reverting to thin-section studies, detailed analysis has been made of the chro-

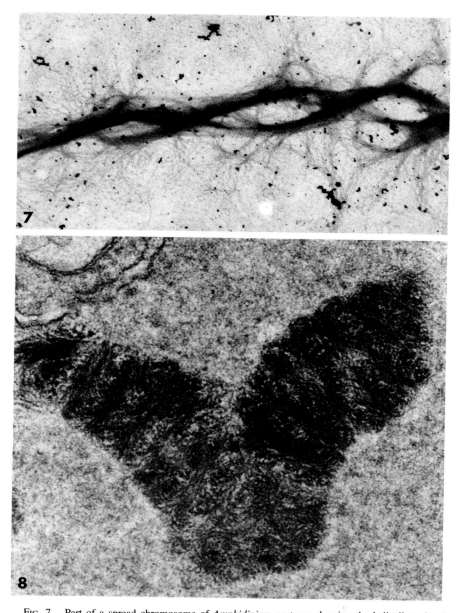

FIG. 7. Part of a spread chromosome of *Amphidinium carterae* showing the helically twisted chromonema and the fine nucleofilaments spreading out from it. Uranyl acetate stain. ×30,000.

FIG. 8. V-shaped chromosome of *Glenodinium hallii* as found at an early stage of mitosis. Preparation as Fig. 5. ×100,000.

mosomes of the freshwater dinoflagellate *Peridinium cinctum* with particular reference to the time of DNA synthesis (Spector *et al.,* 1981; Spector and Triemer, 1981). Here, it was found that there was a brief period in which the tight structure of the chromosome was relaxed and this period corresponds with the maximum synthetic period as evidenced by the uptake of [^{3}H]thymidine. From serial sections of chromosomes at this stage a new model has been proposed for the chromonema. It is thought to consist of a central core of 9-nm fibers around which there is an inner helix and an outer helix, both of 2.5-nm fibrils, to the outer of which are attached a number of 9-nm granules. It would be very interesting to see this model tested by spreading techniques.

Finally, yet another proposal, differing in some respects from all others, has been put forward (Herzog, 1983; Herzog et al., 1984). This model is in large measure based on studies made with nuclei of *prorocentrum* which had been frozen but not fixed chemically and then incubated in various concentrations of EDTA or EGTA, before spreading onto grids. It was found that, in common with some earlier models, there is a double helical bundle (chromonema) consisting of chromosomal fibres. High magnification studies of the fibers revealed a further level of organization down to the DNA double fiber. As noted earlier, it is assumed that the architecture is stabilized by divalent cations. The organizational levels are listed in Table I and illustrated in Fig. 9.

On this basis the whole chromosome should consist only of 4 DNA molecules (Herzog's 6-nm fibers) but this seems far less than are actually observed. The general principle of alternating double helices with single helix supercoiling seems to provide the means of packing the large amount of DNA into the relatively compact chromosomes.

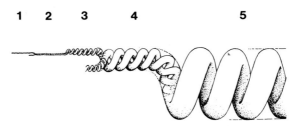

1   2   3   4   5

Fig. 9.   Tentative representation of the helical compaction of dinoflagellate chromosomal DNA in a hierarchy of 6 organization levels. DNA molecules (level 1) form a double helix (level 2) 10 nm thick which is twisted into a single helix 18 nm in diameter (level 3). The fourth level is represented by a double-helically twisted state (25–31 nm thick) of two 18-nm filaments, which in turn are supertwisted into a left-handed single helix (level 5) with a diameter of 43–56 nm. These chromosomal fibers are finally united to form the double helical bundle of the chromosome (not shown here) with a diameter of about 1250 nm. (From Herzog, *et al.,* 1984; reproduced by permission of the authors.)

TABLE I

Chromosome Architecture[a]

Level	Diameter (nm)	Organization	Observed structure
1	6	DNA molecule (double helix)	
2	10	Two twisted DNA molecules	Chromosomal fibers
3	18	Supertwisted 10-nm fibers (single helix)	
4	25–31	Two twisted 18-nm fibers (double helix)	
5	43–56	Supertwisted level 4 fibers (single helix)	Chromonema
6	~1250	Double-helical bundles of level 5 fibers	Chromosome

[a]Adapted from Herzog (1983).

C. The Kinetochore

Light microscopic studies of dinoflagellate mitosis gave the impression that there were no spindle connection regions or kinetochores in the chromosomes (e.g., Dodge, 1963a). The use of electron microscopy changed this to the suggestion that the nuclear envelope was in some way involved in the movement of chromosomes as they appeared to be in contact with the nuclear envelope (Kubai and Ris, 1969). However, it soon became clear that in the parasite *Syndinium* (generally regarded as a dinoflagellate) each of the four chromosomes is attached to the nuclear envelope by disc-like structures which are inserted into it (Ris and Kubai, 1974). The outer side of these structures were attached to spindle microtubules and must therefore be functional kinetochores. Shortly, an analogous structure was reported from another parasite, *Oodinium* (Cachon and Cachon, 1974), and from the free-living dinoflagellate *Amphidinium carterae* (Oakley and Dodge, 1974). In the latter organism, and also other dinoflagellates which have subsequently been investigated (Oakley and Dodge, 1977), the kinetochore was seen simply as a pad of electron-dense material, possibly sitting on the outer side of the nuclear membrane, and becoming conical as it joined the microtubule. It was thought that there was only one microtubule per kinetochore at anaphase and it seemed that there should be two at metaphase (Dodge and Vickermann, 1980). This has recently been confirmed by some elegant electron microscopy on *Amphidinium carterae* (Fig. 10) (Matthys-Rochon, 1983). In this it has been shown that the kinetochores differentiate at the start of nuclear invagination (when spindle tunnels develop) and then microtubules form and connect with them from each pole. After this stage they appear to divide. The appearance of the kinetochores varied with the fixation used. With cacodylate buffered fixatives the kinetochore had a trilaminar form resembling that of the hypermastigine flagellates (Kubai, 1975) but with phosphate buffer they appeared to consist of a nuclear pore of which the outer leaflet is thickened.

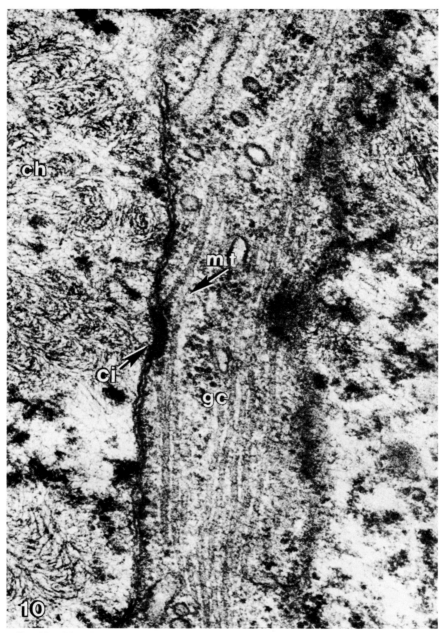

Fig. 10.   Metaphase stage in *Amphidinium carterae* showing a fully formed mitotic tunnel (gc) containing several microtubules (mt) A double kinetochore (ci) is clearly situated on the left side adjacent to a chromosome (ch). ×56,000. (Reproduced by permission of E. Matthys-Rochon.)

V. Chromosome Replication and Division

Several workers have observed V- and Y-shaped chromosomes in dinoflagellate nuclei (Fig. 8) and as these are often very frequent in the early stages of mitosis they have been taken to represent dividing or replicating chromosomes (Leadbeater and Dodge, 1967; Grassé *et al.*, 1965a; Soyer and Haapala, 1974). As noted above, Spector *et al.* (1981) believe that replication takes place with the chromosomes in a relaxed state so, if this is true, the V shapes could then only represent the separation of already replicated chromatids.

There has been much discussion over the time of DNA synthesis. The early suggestion was that, as in prokaryotes, the synthesis was probably continuous (Dodge, 1966). This view has been supported by the autoradiographic studies of Filfilan and Sigee (1977). However, other workers using a variety of techniques on a range of dinoflagellates appear to have shown conclusively that there is a distinct S phase during the interphase when most DNA synthesis takes place (Franker, 1972; Franker *et al.*, 1974; Loeblich, 1976). Perhaps the difference between the two views is not as serious as at first appears since it seems that although most of the synthesis takes place (isotope is taken up) during one period of the cell cycle there is a residual uptake for much of the rest of the cycle (Galleron and Durrand, 1979). In one study of the chromosomes of a *Gyrodinium* species (Ishio *et al.*, 1978) it has been suggested, on the basis of TEM observations, that chromosome diameter increases fourfold during the period after late anaphase by the incorporation of nucleogranules which autoradiography showed were synthesized at the margins of the chromosomes. It is difficult to reconcile this work with that reported above.

VI. Conclusions

It is clear that dinoflagellate chromosomes are unique structures. Structurally they probably have more in common with nuclei of bacteria but in terms of their behavior (i.e., mitosis) they show similarities with normal eukaryotic chromosomes. So little is known about their genetic capabilities that it is impossible at this stage to make postulates about their affinities in this respect. In short, while there is very much to be discovered about these structures, it seems clear that they provide either a fascinating bridge between the nuclear arrangements in prokaryotes and those in eukaryotes or, alternatively, a quite remarkable and successful experiment which, like the dinosaurs, had only limited evolutionary potential.

REFERENCES

Allen, J. R., Roberts, T. M., Loeblich, A. R., III, and Klotz, L. C. (1975). *Cell* **6,** 161–169.
Borgert, A. (1910). *Arch. Protistenkd.* **20,** 1–46.

Bouligand, Y., Soyer, M.-O., and Puiseux-Dao, S. (1968). *Chromosoma* **24**, 251–287.

Cachon, J., and Cachon, M. (1974). *C. R. Hebd. Seances. Sci. Acad. Paris* **278**, 1735–1737.

Dodge, J. D. (1963a). *Arch. Protistenkd.* **106**, 442–452.

Dodge, J. D. (1963b). *Bot. Mar.* **5**, 121–127.

Dodge, J. D. (1964). *Arch. Microbiol.* **48**, 66–80.

Dodge, J. D. (1966). *In* "The Chromosomes of the Algae" (M. B. E. Godward, ed.), pp. 95–115. Arnold, London.

Dodge, J. D., and Vickerman, K. (1980). *Soc. Gen. Microbiol., Symp.* **30**, 77–102.

Entz, G. (1921). *Arch. Protistenkd.* **43**, 415–430.

Filfilan, S. A., and Sigee, D. C. (1977). *J. Cell Sci.* **27**, 81–90.

Fine, K. E., and Loeblich, A. R., III (1976). *Proc. Biol. Soc. Washington* **89**, 275–288.

Franker, C. K. (1972). *J. Phycol.* **8**, 264–268.

Franker, C. K., Sakhrani, L. M., Pritchard, C. D., and Lamden, C. A. (1974). *J. Phycol.* **10**, 91–94.

Galleron, C., and Durrand, A. M. (1979). *Protoplasma* **100**, 155–165.

Gavrila, L. (1977). *Caryologia* **30**, 273–287.

Giesbrecht, P. (1961). *Zentralbl. Bakteriol. Parasitenkd. Abt. I* **183**, 1–37.

Giesbrecht, P. (1965). *Zentralbl. Bakteriol. Parasitenkd. Abt. I* **196**, 516–519.

Grassé, P. P., and Dragesco, J. (1957). *C. R. Hebd. Seances Acad. Sci. Paris* **245**, 2447–2451.

Grassé, P. P., Hollande, A., Cachon, J., and Cachon-Emjumet, M. (1965a). *C. R. Hebd. Seances Acad. Sci. Paris* **260**, 1743–1747.

Grassé, P. P., Hollande, A., Cachon, J., and Cachon-Enjumet, M. (1965b). *C. R. Hebd. Seances Acad. Sci. Paris* **260**, 6975–6978.

Grell, K. G., and Wohlfarth-Bottermann, K. E. (1957). *Z. Zellforsch. Mikrosk. Anat.* **47**, 7–17.

Haapala, O. K., and Soyer, M.-O. (1973). *Nature (London)* **137**, 195–197.

Haapala, O. K., and Soyer, M.-O. (1974). *Heriditas* **78**, 146–150.

Hall, R. P. (1925). *Univ. Calif. Publ. Zool.* **28**, 29–64.

Hamkala, B. A., and Rattner, J. B. (1977). *Chromosoma* **60**, 39–47.

Herzog, M. (1983). *Doctoral thesis University of Paris.*

Herzog, M., and Soyer, M.-O. (1980). *Cell Biol.* **23**, 295–302.

Herzog, M., and Soyer, M.-O. (1982a). *Cell Biol.* **27**, 151–155.

Herzog, M., and Soyer, M.-O. (1982b). *Cell Biol.* **30**, 33–41

Herzog, M., and Soyer, M.-O. (1983). *Eur. J. Cell Biol.* **30**, 33–41.

Herzog, M., Boletsky, S., and Soyer, M.-O. (1984). *Origins of Life* **13**, 205–215.

Hinnebusch, A. G., Klotz, L. C., Immergut, E., and Loeblich, A. R., III (1980). *Biochemistry* **19**, 1744–1755.

Hollande, A. (1974). *Protistologia* **10**, 413–451.

Holt, J. R., and Pfiester, L. A. (1982). *Am. J. Bot.* **69**, 1165–1168.

Ishio, S., Chen, C. C., and Yano, T. (1978). *Bull. Jpn. Sci. Fish.* **44**, 185–192.

Kearns, L. P., and Sigee, D. C. (1980). *J. Cell Sci.* **46**, 113–127.

Kubai, D. F. (1975). *Int. Rev. Cytol.* **43**, 167–227.

Kubai, D. F., and Ris, H. (1969). *J. Cell Biol.* **40**, 508–528.

Leadbeater, B. S. C. (1967). Ph.D. thesis, University of London.

Leadbeater, B., and Dodge, J. D. (1967). *Arch. Mikrobiol.* **57**, 239–254.

Livolant, F. (1978). *Chromosoma* **68**, 45–58.

Livolant, F., and Bouligand, Y. (1978). *Chromosoma* **68**, 21–44.

Livolant, F., and Bouligand, Y. (1980). *Chromosoma* **80**, 97–118.

Loeblich, A. R., III (1976). *Stadler Symp.* **8**, 111–128.

Loeblich, A. R., III, Schmidt, R. J., and Sherley, J. L. (1981). *J. Plankton Res.* **3**, 67–79.

Loper, C. L., Steidinger, K. A., and Walker, L. M. (1980). *Trans. Am. Microsc. Soc.* **99**, 343–346.

Matthys-Rochon, E. (1983). Doctoral thesis, University of Paris.

Oakley, B. R., and Dodge, J. D. (1974). *J. Cell Biol.* **63,** 322–325.

Oakley, B. R., and Dodge, J. D. (1977). *Cytobios* **17,** 35–46.

Oakley, B. R., and Dodge, J. D., (1979). *Chromosoma* **70,** 277–291.

Rae, P. M. M. (1973). *Proc. Natl. Acad. Sci. U.S.A.* **70,** 1141–1145.

Rae, P. M. M. (1976). *Science* **194,** 1062–1064.

Ris, H. (1962). *Symp. Int. Soc. Cell Biol.* **1,** 69–88.

Ris, H., and Kubai, D. F. (1974). *J. Cell Biol.* **60,** 702–720.

Rizzo, P. J., and Burghardt, R. C. (1980). *Chromosoma* **76,** 91–99.

Rizzo, P. J., and Burghardt, R. C. (1982). *BioSystems* **15,** 27–34.

Rizzo, P. J., and Nooden, L. D. (1972). *Science* **176,** 796–797.

Rizzo, P. J., and Nooden, L. D. (1974). *Biochim. Biophys. Acta* **349,** 402–414.

Roberts, T. M., Tuttle, R. C., Allen, J. R., Loeblich, A. R., III, and Klotz, L. C. (1974). *Nature (London)* **248,** 446–447.

Sarma, Y. S. R. K., and Shyam, R. (1974). *Br Phycol. J.* **9,** 21–29.

Shyam, R., and Sarma, Y. S. R. K. (1978). *Bot. J. Linn. Soc.* **76,** 145–159.

Sigee, D. C. (1982). *Protoplasma* **110,** 112–120.

Sigee, D. C. (1983). *Bot. J. Linn. Soc.* **88,** 127–147.

Sigee, D. C., and Kearns, L. P. (1981a). *Protoplasma* **105,** 213–223.

Sigee, D. C., and Kearns, L. P. (1981b). *Tissue Cell* **13,** 441–451.

Sigee, D. C., and Kearns, L. P. (1981c). *Cytobios* **31,** 49–65.

Sigee, D. C., and Kearns, L. P. (1982). *Cytobios* **33,** 51–64.

Soyer, M. O., and Haapala, O. K. (1974). *J. Microsc.* **19,** 137–146.

Spector, D. L., and Triemer, R. E. (1981). *BioSystems* **14,** 289–298.

Spector, D. L., Vasconcelos, A. C., and Triemer, R. E. (1981). *Protoplasma* **105,** 185–194.

Stosch, H. A., von (1973). *Br. Phycol. J.* **8,** 105–134.

(or superbeads). Between such supernucleosomes, rather extended material can occur, so that individual superbeads are visible in ultrathin sections or spread chromatin (for models see Worcel, 1977; Scheer and Zentgraf, 1978; Erenpreiss *et al.,* 1981).

C. Higher Order Structures

How the higher order packaging of DN$ into mitotic chromosomes is achieved is not yet clear, and several models have been put forward: the folded-fiber model (e.g., DuPraw, 1970), the coiling hierarchy model (e.g., Sedat and Manuelidis, 1977), the "unit fiber" (tube) model (e.g., Bak and Zeuthen, 1977), and the radial loop model (e.g., Marsden and Laemmli, 1979). This problem will, however, not be discussed in this article (for discussion, see Nagl, 1982b).

Chromatin condensation at interphase evidently follows a different pathway, which is mainly characterized by cross-linking of chromatin fibers, but discrete thermal transitions of chromatin and melting enthalpies revealed the existence of specific tertiary, quaternary, and quinternary structures of chromatin DNA (Nicolini *et al.,* 1983). With respect to the complete nucleus, interphase chromatin is organized into loops, domains, and areas occupied by individual chromosomes (see below). Talking about condensed chromatin requires, however, some terminological clearance, because this issue covers a variety of chromatin moieties with different molecular and biological characteristics. Unfortunately, no clear terminology has been accepted so far by all scientists.

Condensed chromatin denotes any kind of chromatin that occurs in the form of dense patches at the light or electron microscope level, irrespective of the cause of condensation or its functional significance.

Condensed euchromatin is a chromatin fraction which rests on different mechanisms in plants and animals. In plants, its amount is fairly species-specific (Nagl, 1979a, 1982a). This chromatin appears in the form of chromomeres or chromonemata and can be distinguished as euchromatin in the light microscope only, while it shows the same electron density as the heterochromatin in the electron microscope (see Section II, E). In higher animals, on the other hand, a tissue-specific amount and pattern of condensed euchromatin can be found, caused by chromosomal proteins and their modification, by ions etc. Here chromatin condensation evidently represents the deterministic step of tissue-specific gene inactivation; often this type of chromatin is uncritically called *heterochromatin,* but in this case, at least the addition "functional" should be given (for detailed discussion see Nagl, 1979b, 1982a).

Heterochromatin sensu strictu includes constitutive heterochromatin, i.e., chromatin that is always mitotically condensed at interphase (except in short periods of the cell cycle), because of its DNA base sequences and the chromosomal proteins bound to these sequences. Normally, it is rich in highly re-

petitive DNA (such as satellite DNA), late replicating, poor in genes, and—at
the light microscope level—C-banding positive. In female mammals, the major
part of one of the two X chromosomes behaves like heterochromatin, although its
nucleotide sequence is the same as in the decondensed X chromosome; for this
material, the term *facultative heterochromatin* or sex chromatin was introduced
(Smith, 1945).

Although heterochromatin is thought to remain in a mitotically condensed
state, no clear higher order structure could be detected by three-dimensional (3D)
reconstruction of serial sections of chromocenters (Müller and Nagl, 1983),
probably due to additional cross-linking of chromatin domains.

As condensed euchromatin is species-specific (karyotypical) in plants, but
developmental- and tissue-specific (functional) in animals, it can be assumed that
the mechanisms underlying condensation are different, too. Actually, it seems
that chromonemata and chromomeres in plant nuclei are relics of mitotic condensation, while clumps of condensed euchromatin in animal nuclei are the result of
the organization of chromatin fibers by nonhistone chromosomal proteins into
loops and domains. However, it was also suggested that the domain is the basis
of longitudinal differentiation of metaphase chromosomes as revealed by Giemsa
banding (Comings, 1978). Besides nonhistone chromosomal proteins and histone (particularly H1) modifications, various other structural components of the
nucleus may have a role, such as the nuclear matrix (Small *et al.,* 1982), or the
nuclear envelope (especially in the case of peripheral condensation pattern; see
the ''drapery model'' of Nicolini, 1979), or a heterochromatin scaffold (Erenpreisa, 1982). The possible regulation of chromatin condensation is discussed in
Section VI.

In functional terms, the pattern of condensed chromatin reflects the tissue-
specific pattern of inactivated chromatin domains and genes included within this
domains (see Fig. 16 in Nagl, 1982a). This interpretation holds, however, only if
the individual chromosomes own a fixed position within the interphase nucleus.

D. CHROMOSOME DISPOSITION AT INTERPHASE

There is increasing evidence by both analysis of chromosome arrangement at
metaphase and analysis of Giemsa-stained interphase nuclei (and other methods)
that there exists an ordered spatial arrangement of genetic material within the
interphase nucleus (e.g., Bennett, 1982; Cremer *et al.,* 1982; Tanaka, 1981;
Antoine *et al.,* 1982; Avivi *et al.,* 1982a,b; Evans and Filion, 1982; Hager *et al.,*
1982).

The mechanisms by which the discrete arrangement of chromosomes at interphase is achieved may be manifold. The position could be the result of mitotic
movement of chromosomes, of the attachment of chromosomes (e.g., telomeres)
to the nuclear envelope, or of the association of chromosome arms of similar size

(and hence DNA content, and possibily similar DNA sequences and related genes). The result is a highly ordered disposition of chromosomes and genes at interphase.

E. Nuclear Ultrastructure and Structure

While the visualization of nucleosomes normally requires the artificial destruction of the native structure of chromatin, ultrathin or thick sections of nuclei as well as squash preparations reveal the actual arrangement of dense and diffuse chromatin within a nucleus. It must be mentioned that patterns obtained in these ways may give an impression which is different from the 3D image obtained from isolated, complete nuclei as they are used, for instance, in flow cytofluorometric studies.

On the basis of nuclear images as seen in the light and electron microscope, a number of types of nuclear organization can be classified. Here only the situation in angiosperms will briefly be reviewed (for a more general discussion see Tschermak-Woess, 1963; Nagl, 1979a, 1982a,b).

Species-, genus-, and even family-specific nuclear structures have been observed in higher plants for 50 years. The main types, arranged according to the proportion of condensed chromatin and the structural complexity of the nuclei, are the following.

Diffuse (with respect to euchromatin) or *chromocentric nuclei* (with respect to heterochromatin). Their euchromatin is completely diffuse, nearly invisible, while the constitutive heterochromatin forms dense blocks, so-called chromocenters (Fig. 3a).

Chromomeric nuclei. In this type of nuclei, part of the euchromatin is in a condensed state, forming small granular patches (Fig. 3b). It must be emphasized that the number and size of interphase chromomeres do not necessarily correspond to the number and size of meiotic (pachytene) chromomeres, as the chromomere is a dynamic structure of chromosome organization. The appearance of chromomeres is based on the DNA base sequence, thus reflecting genetic material somewhat in between of constitutive heterochromatin and diffuse euchromatin. Chromomeric nuclei may, in addition, possess chromocenters.

Chromonematic nuclei. In nuclei of this type most, if not all, of the euchromatin is condensed into thread-like structures of about 0.2 to 0.5 μm in diameter (Fig. 3c). Chromocenters may be there, in addition. According to another terminology, nuclei of this type are named "reticulate" (Lafontaine, 1974; Barlow, 1977).

Of course, classification of nuclear structure is man-made, and in nature all kinds of transition forms, intermediate and composed structures can be found. In general, the complexity of nuclear structure apparently depends on the complex-

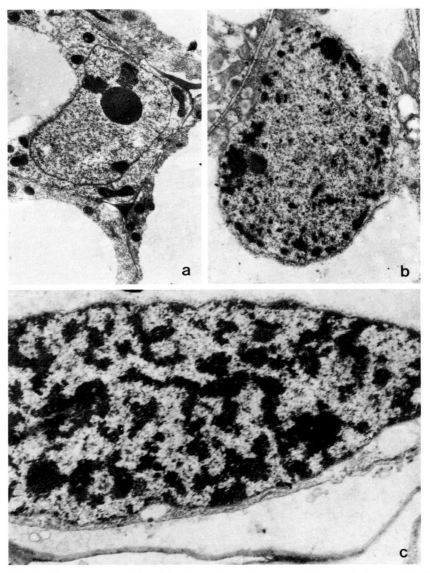

Fɪɢ. 3. Electron micrographs of the three characteristic structural types of plant nuclei: (a) diffuse/chromocentric nucleus of *Cucumis sativus*, (b) chromomeric nucleus of *Senecio viscosus*, (c) chromonematic nucleus of *Allium flavum;* ×6500.

ity of DNA base composition as visible in a derivative melting profile of DNA (Capesius and Nagl, 1978; Nagl, 1982a).

In spite of the high species specificity of nuclear structures in plants, some variation can be observed (for details see Section III,D). Due to fusion of chromocenters to collective chromocenters, or their disaggregation into heterochromomeres, a moderate tissue-specific pattern can arise (e.g., Hirahara, 1980; Tanaka, 1982).

III. Determination of Chromatin Organization in Plants

If nuclear structures in plants are species-specific, then there must exist some species-specific determinants. This is actually the case.

A. NUCLEOTYPIC FACTORS

As "nucleotype" the DNA mass of a nucleus is described (2C value), independent of its genetic information (Bennett, 1973). During recent years, a number of nucleotypic effects have been detected. For instance, the 2C value exerts a dominant effect on the duration of the mitotic and meiotic cell cycle, the minimum generation time, nuclear and cell size, DNA sequence intersperion pattern, and chromosome disposition at interphase (for reviews see Bennett, 1973, 1982; Nagl, 1976a, 1978). Also the species-specific nuclear structure appears to be the result of the nucleotype, although the amount frequency, complexity, and distribution of repetitive DNA may be of influence, too (Nagl, 1979a; Nagl *et al.*, 1983b). In general, DNA-poor species exhibit diffuse/chromocentric nuclei, and DNA-rich species chromonematic nuclei. Those species with intermediate 2C values own chromomeric and transition form nuclei (Fig. 4).

The DNA content does not only influence chromatin organization in qualitative terms, but also in quantitative ones: The proportion of condensed chromatin (as characteristic for a species) is closely correlated with the 2C value, although the regression slope is different in DNA-poor and DNA-rich discots, and in monocots. Evidently it is family- or genus-specific (Nagl, 1982a,b). In some taxa, a correlation coefficient of 1, or close to 1, has been calculated from the measurements (Nagl and Bachmann, 1980; Nagl, 1980).

An interesting nucleotypic effect is that on cell size, which is known as the "nucleocytoplasmic relationship" since the first days of cytology (for a recent discussion see Cavalier-Smith, 1982). This clearly shows that the non-protein-coding DNA, and this is the huge mass of the genome, does have effects and must, therefore, not be designated as junk. Moreover, the question arises, how (in physiological terms) noncoding DNA can influence cell volume and control it

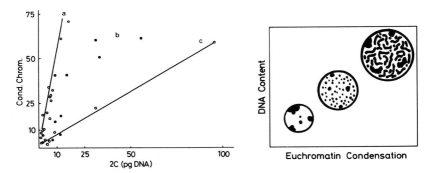

FIG. 4.   The relationship between the 2C nuclear DNA content and the species-specific proportion of condensed chromatin in angiosperm nuclei. Left: some actual measurements in DNA-poor
dicots (a), monocots (b), and DNA-rich dicots (c); modified from Nagl (1982a). Right: images of
nuclei with different 2C value; modified from Nagl (1982b).

very exactly. The suggestion of Nagl and Popp (1983) and Popp and Nagl (1983)
that electromagnetic properties of DNA play in this respect (as in others) a
dominant role represents the only possible explanation so far available (see also
Section VI,C).

B. GENETIC FACTORS

While nucleotypic effects exerted by the DNA content (2C value) and/or the
amount of certain DNA fractions (repetitive sequences) apparently play the major role in determining the gross organization of plant cell nuclei, genetic factors
cannot be ruled out. Unfortunately, our knowledge on this topic is rather limited.
Direct evidence comes from a mammal. Goldowitz and Mullen (1982) reported
that nuclear morphology deals as a cell marker in chimeric brain of ichthyosis
(hair-deficient) mutant mice. Our own unpublished results obtained in cultured
tobacco cell lines give more indirect evidence. Suspension cultures of wild-type
cells of *Nicotiana tabacum* L. cv. Xanthi (TX1) exhibit the characteristic, species-specific nuclear ultrastructure. However, the nuclei of a *p*-fluorophenylala-
nine-resistant cell line, TX4, which shows high phenylalanine ammonialyase
(PAL) activity and accumulation of cinnamoyl putrescines (Berlin *et al.*, 1982),
exhibit a totally diffuse chromatin organization (Fig. 5; Nagl *et al.*, 1984).

C. STABILITY OF NUCLEAR ORGANIZATION

All the nuclei of mature cells of a given plant species exhibit the same amount
of condensed chromatin (Fig. 6a), except some highly specialized cells such as
antipodal cells or microsporocytes (Nagl *et al.*, 1983a). The nuclear structure

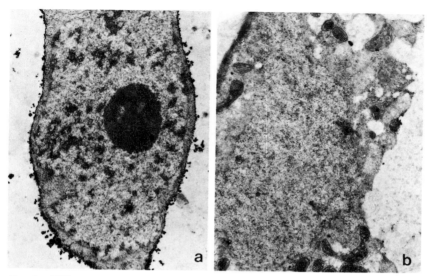

Fᴵɢ. 5. Differences in chromatin organization in two cell lines of *Nicotiana tabacum* (suspension cultures): (a) line TX1 with the species-specific chromomeric nuclei, (b) line TX4 (phenylalanine resistant) with diffuse nuclei; ×4100.

also shows thermostability at least at growth temperatures between 10 and 40°C, and under the influence of inhibitors of RNA synthesis (Fig. 6b; Nagl, *et al.,* 1983a). This is in contrast to mammalian nuclei, which respond with prominent changes in chromatin condensation under those influences (e.g., Brasch and Sinclair, 1978; Derenzini *et al.,* 1981). However, it cannot be excluded that changes do occur which are too small to be detected by electron microscopic stereology. One has to consider that only 1 to 0.1%, or even less, of plant DNA represents genes *sensu strictu,* so that nearly the entire DNA is composed of noncoding single copy and repetitive sequences (Nagl, 1983b; Nagl *et al.,* 1983b).

The structural stability of chromatin is also expressed by the fact that no differences occur in the electrophoretic mobility of histones in the meristem, elongation zone, and differentiated tissue of roots (for *Zea mays* see Burkhanova *et al.,* 1975; Nichlisch *et al.,* 1976).

D. Cʜᴀɴɢᴇꜱ ᴏꜰ Cʜʀᴏᴍᴀᴛɪɴ Cᴏɴᴅᴇɴꜱᴀᴛɪᴏɴ ᴅᴜʀɪɴɢ Dɪꜰꜰᴇʀᴇɴᴛɪᴀᴛɪᴏɴ

Changes in nuclear size in differentiating plant tissues are commonly related to somatic polyploidization (see the following section). However, nuclear size may also vary independently of the DNA content (reviewed by Nagl, 1978). This is evidently due to various degrees of hydration and, particularly, various amounts

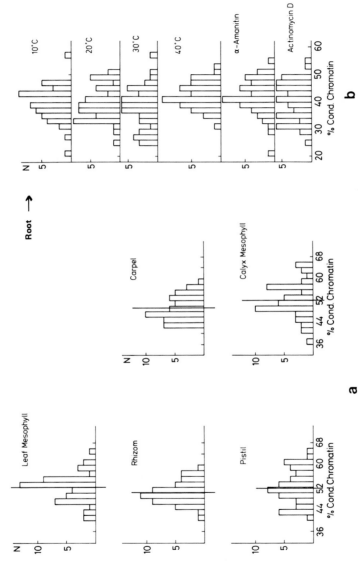

Fig. 6. Histograms showing the stability of the percentage of condensed chromatin in a plant, *Tradescantia virginiana*: (a) nuclei of mature cells of various tissues; (b) nuclei of root tips (noncycling cells) grown at different temperatures and within solutions of inhibitors or RNA synthesis (modified from Nagl *et al.*, 1983a).

of protein present in the nuclei. This was, for instance, demonstrated in differentiating stem cells of *Pisum sativum* (Mitchell and van der Ploeg, 1982). An increase in nonhistone chromosomal proteins has been observed and interpreted as a factor in the establishment of tissue-specific RNA synthesis (Burkhanova *et al.*, 1975). On the other hand, chromatin template activity is known to be highest in proliferating cells, and indeed Geurri *et al.* (1982) described increased amounts of some high-molecular-weight nonhistone chromosomal proteins during an induction period to cell division in tobacco tissue cultures. Minor differences in the histone fractions of meristematic and differentiating tissues are apparently rather related to chromatin changes during the cell cycle (Nicklisch *et al.*, 1976) (see Section V,D).

During germination and seedling growth of wheat, conformational changes of chromatin were found by means of DNase II digestion and polylysine binding capacity essays (Sugita and Sasaki, 1982), which may be related to changes in transcription. Dhillon and Miksche (1982, 1983) described a decrease in heterochromatin (condensed euchromatin?) in germinating roots of peanut and soybean, and interpreted it as an expression of increased gene activity. Similar observations were reported for organ differentiation in pea (Voges and Fellenberg, 1976) and xylem vessels (Lai and Srivastava, 1976). The changes found in the root apex of *Zea mays* on the basis of electron micrographs (Barlow and Barlow, 1980; Barlow and Hines, 1980) may be related to differences in mitotic activity and senescence, respectively (see below). Changes in the area fraction of nucleoplasm occupied by chromocenters (heterochromatin?) were also described during early stages of differentiation in roots of the water-fern, *Azolla* (Barlow *et al.*, 1982). But also in this case it is not clear how much of these changes are due to difference in the mitotic activity (cycling cells always exhibit a higher average proportion of condensed chromatin than noncycling cells) and differences in nuclear protein content (and volume). The sense of a relation of changes in the fraction of heterochromatin to nuclear activity is, in general, somewhat questionable, because this fraction very likely does not include genes.

Clear changes in the proportion of condensed chromatin take place during the transition to flowering (Jacqmard *et al.*, 1972). Havelange and Bernier (1974) could show that, in the evocked apical meristem of mustard, this change occurs due to an increase of the area (volume) occupied by dispersed chromatin, while the area (volume) occupied by condensed chromatin remained constant. This indicates that there does a dispersion of condensed chromatin not takes place, but a further diffusion of the already decondensed fraction. This is consistent with findings in animals, where hormone treatment cause the further decondensation of diffuse chromatin, while the tissue-specifically condensed fraction is apparently irreversibly condensed (Kiefer *et al.*, 1974). The prominent increase in nucleolar size during floral evocation may be the result of both, increase in ribosomal gene activity and ribosomal gene amplification (Jacqmard *et al.*,

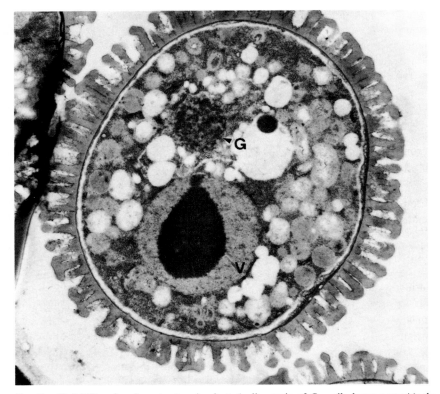

Fig. 7.   Variability of nuclear structure in plants (pollen grain of *Capsella bursa-pastoris*); the generative nucleus (G) is highly condensed, while the vegetative nucleus (V) is highly decondensed and displays a large nucleolus; ×2000.

1981). The transition to flowering is also accompanied by qualitative changes in protein synthesis (Pierard *et al.*, 1980). No changes in the chromatin were found between the noninduced and the induced apical meristem of *Xanthium* (Havelange, 1980), indicating that data are still not sufficient for a generalization of the changes observed in some species.

Regularly, dramatic deviations from the species-specific chromatin organization are found in cells of the gametophyte. For instance, the vegetative nucleus of the pollen grain is always larger than that of the generative nucleus, and it exhibits completely decondensed chromatin and a large nucleolus, while the latter is more condensed (Fig. 7). In the ovule, endosperm nuclei, antipodal nuclei, and others often display an altered, normally decondensed structure (Nagl *et al.*, 1983a).

Senescence and cell death is an integral part of plant development and deals, at least in certain tissues, the recycling of cellular material (nutrients, hormones) as, for instance, during embryogenesis (Nagl, 1976b; Gärtner and Nagl, 1980; Barlow, 1982). Senescence of plant cells is apparently always accompanied or even induced by loss of nuclear DNA (Dhillon and Miksche, 1981; Drumm and Nagl, 1982; Hesemann and Schröder, 1982). The loss of DNA is reflected by alterations in nuclear structure, which finally lead to an increase in the percentage of condensed chromatin, due to loss of diffuse euchromatin and nuclear shrinkage (Drumm and Nagl, 1982).

E. POLYPLOID AND POLYTENE NUCLEI

Somatic polyploidy occurs very frequently among higher plants (reviewed by Nagl, 1978). Most angiosperms show a tissue-specific pattern of various levels of endopolyploidy (i.e., polyploidy due to the two types of endocycles, the endomitotic and the endoreduplication cycle) which comprises up to 71% of all nonmeristematic cells (e.g., Frisch and Nagl, 1979). The highest levels of endopolyploidy are normally found in nutritive cells of the ovule (such as embryo-suspensor cells, endosperm haustoria). In some cases highly endopolyploid nuclei display polytene chromosomes, structures that originate if no mitotic condensation process interrupts polyploidization (reviewed by Nagl, 1978, 1981). In spite of various structures of polyploid nuclei which deviate from the characteristic image of diploid cells, the gross amount of condensed chromatin remains the same. However, minor changes become visible due to the increased size of nuclei and chromatin structures. Examples are local or general puffing phenomena in polytene nuclei (review: Nagl, 1974). On the other hand, caution is necessary in the interpretation of structural changes in terms of transcriptional activity. Some of them are more related to differential DNA replication (DNA amplification and underreplication) than to RNA synthesis (Nagl, 1972, 1979b).

IV. Chromatin Organization and Differentiation in Mammals

In this section, the main differences in plant nuclei as they are found in mammalian nuclei (and those of other vertebrates) will briefly be summarized. These differences may be primarily related to the fact that most plant cells are totipotent throughout their life, while mammalian cells become irreversibly ''determined'' early in development. Determination is evidently controlled via the pattern of chromatin condensation, thus limiting the genes which remain available for fine-tuning regulation and hence cell differentiation. This is consistent with the suggestion of other authors that the structural organization of chromatin in vertebrate cells determines the irreversible restrictions in gene expression specific for differentiation (Tsanev, 1974; Nicolini, 1979, 1980a,b). Escape of the tissue-specific determination and differentiation in the case of malignant transformation and carcinogenesis is always accompanied by a change in chromatin organization (see Section V,E).

A. CHANGES IN NUCLEOSOME REPEAT LENGTH

All higher organisms which have been examined so far appear to have 140 base pairs of DNA associated with the nucleosome core particle. However, the DNA repeat length of chromatin has been found to vary due to differences in the length of linker DNA between core particles (Kornberg, 1977; Felsenfeld, 1978). While most animal tissues exhibit DNA repeat lengths close to 200 base pairs, some deviations have been found in connection to transcriptional activity and development (no comparative studies do exist for plant tissues).

A short chromatin repeat length (often the term "DNA repeat length" is used, which is confusing with the length unit of repetitive DNA sequences) of 160 base pairs has been found in differentiating neutronal nuclei (Thomas and Thompson, 1977; Brown, 1982), that is evidently caused by an exchange of some of the histones of the nucleosome core (Brown, 1980). Nucleosome repeat length alternations are also associated with the *in vitro* differentiation of murine embryonal carcinoma cells to extraembryonic endoderm; in this case, changes of histone H1 forms could be detected (Oshima *et al.*, 1980). This is understandable, as histone H1 subfractions apparently have an important role in the folding of chromatin into certain higher order structures (Biard-Roche *et al.*, 1982).

In this connection it must be emphasized that nucleosomal chromatin organization is not as much independent of the DNA sequence as generally believed (see Maio *et al.*, 1977; Zhurkin, 1982). Also the higher order structures, such as supercoiled loops, show a relationship of DNA sequences to organizing systems, such as the nuclear matrix (Small *et al.*, 1982).

B. FUNCTIONAL PATTERNS OF CHROMATIN CONDENSATION

In contrast to plants, higher animals display patterns of condensed chromatin, which are not species-specific but specific for distinct stages of development, specific tissues and cells, and related to template and gene activity. In general, embryonic cells and permanently proliferating stem cells exhibit a diffuse chromatin structure, while certain portions of the chromatin condense during determination and differentiation, as shown above. It seems that large portions of the genome become irreversibly silenced. This structural type of inactivation represents the dominant step in the control of tissue-specific transcriptional capacity and selectivity.

The close relationship between chromatin decondensation and transcription, and chromatin condensation and inhibition of transcription, was repeatedly shown by experiments using autoradiography, treatment with ions, hormones, and inhibitors of RNA synthesis (e.g., Hsu, 1962; Kiefer *et al.*, 1974; Brasch and Sinclair, 1978; Derenzini *et al.*, 1981; Paiement and Bendayan, 1982). The decondensation of chromatin in diploid nuclei can thus be compared to the puffing of bands (i.e., chromomeres) of polytene chromosomes in insect larvae. Other studies, however, indicate that changes in chromatin pattern are also related to posttranscriptional activities (Puvion and Viron, 1981). Good examples for the transformation of nuclear morphology during maturation are the various types of mammalian leukocytes, which finally exhibit well-known specific forms and condensation patterns of chromatin, the different cell types of the retina of the guinea pig eye (Schmalenberger and Nagl, 1979; Schmalenberger, 1980), and the cartilaginous cells of the newt (Jeanny and Gontcharoff, 1978, 1981). In the latter case it was shown that condensation is related to aging, while decondensation occurs during regeneration and restart of proliferation. Also in the human placenta, the embryonic trophoblast nuclei undergo subsequent condensation during pregnancy, and many of them are totally condensed (pycnotic) in the term placenta which is ready for parturition (W. Nagl and B. Fuhrmann, unpublished). These events indicate again that the developmental program of a cell starts, after proliferation has stopped, with continuous condensation of chromatin into a specific pattern, and proceeds on until senescence and cell death. In parallel, transcription is first qualitatively limited to tissue-specific genes, and later it is quantitatively reduced until the cell is no longer able to fulfill its functions.

V. Active Genes and Their Chromatin Environment

As discussed in previous sections, condensation of the chromatin domain within which a gene is embedded can be seen as the limiting factor for its activation. Dependent on the structure of the chromatin surrounding a gene, itself and its recognition and promoter sequences are exposed to, or hidden before regulatory proteins and the enzymatic machinery of transcription. In addition, however, the lower levels of chromatin organization of an actually transcribed gene have been found to be different from those of inactive genes.

A. CHROMATIN STRUCTURE OF ACTIVE GENES

DNase I (i.e., pancreatic DNase, EC 3.1.4.5) is an endonuclease with little DNA-sequence specificity. When eukaryotic nuclei are partially digested with this enzyme, the resultant DNA fragments form a continuum of sizes, causing a uniform smear on agarose gels. Weintraub and Groudine found in 1976 that, however, active chromatin is preferentially digested. Many experiments which have been done since then indicate that a region of the chromatin, including the DNA sequence of transcription and extending many kilobases beyond it, is very sensitive to DNase I.

More recently, Wu *et al.*, (1979) detected a DNase I hypersensitive site at or near the 5' end of genes and suggested that the altered chromatin conformation at this site is necessary, but not sufficient, for transcription by RNA polymerase II. Wu (1982) reported that this exposed structure at the 5' terminus of several heat-inducible genes in *Drosophila* is present before and during induction. As in other genes (see minireview by Elgin, 1981), the DNase I-hypersensitive site can be equated with the upstream sequences (about 300–400 nucleotide pairs), sometimes extending through the TATA box. Hence, gene activation evidently starts with alteration of chromatin structure at the DNase I-hypersensitive site (see also Lohr, 1983; Tsaneu, 1983).

Moreover, there is evidence that chromatin decondensation precedes transcription and remains somewhat longer than RNA synthesis can be detected (e.g., Rindt and Nover, 1980; Elgin, 1981; Sledziewski and Young, 1982; Parslow and Granner, 1982).

Somewhat controversial is the question of whether active genes are organized, in principle, within nucleosomal chromatin or not. It seems to be evident that genes for messenger RNA are in part organized into nucleosomes, the number of which depends on the degree of activity (Chambon, 1977; Scheer, 1978). Nucleosomes were found very close to transcriptional complexes in maize (Greimers and Deltour, 1981), while no nucleosomes were found in very active lampbrush chromosomes (Scheer, 1978). The situation of active ribosomal genes is not as clear. Electron micrographs of spread chromatin indicate the absence of nucleosomes (Franke *et al.*, 1976; Labhart and Koller, 1982), while gel electrophoresis of DNA from nuclei digested with *Micrococcus* nuclease indicate the presence of a nucleosomal organization in plants (Leber and Hemleben, 1979; Leweke and Hemleben, 1982). Figure 8 shows the hybridization of DNA fragments from micrococcal nuclease digest to total *Vicia faba* chromatin (i.e., DNA of nucleosomal size) to cloned plasmid pTA 250 containing wheat ribosomal DNA (constructed by Gerlach and Bedbroock, 1979). Although this gel indicates the nucleosomal organization of ribosomal DNA, it cannot be deduced from this result that the *active* genes are represented, because most of the ribosomal genes are in an inactive state within the nucleolus organizer heterochromatin.

In general, evidence exists that the transcriptional activation of a gene is accompanied, or better preceded, by chromatin conformational changes. It is very likely that nonhistone chromosomal proteins are involved in the specific open configuration of chromatin around active genes (reviewed by Cartwright *et al.*, 1982). Nevertheless, these proteins are surely not the basic control elements in chromatin structure (see Section VI).

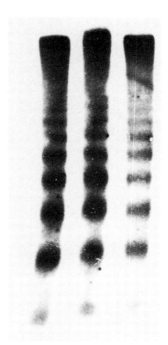

FIG. 8. Nucleosomal organization of ribosomal DNA from *Vicia faba* root tips. Filter hybridization (Southern blot) of nucleosomal DNA gels with ^{32}P-labeled, cloned plasmid DNA containing ribosomal DNA of wheat.

B. TRANSCRIPTION OF REPETITIVE DNA AND CONDENSED CHROMATIN

Recently, attention has been paid to the transcription of repetitive DNA (Davidson and Britten, 1979). Several reports have shown that, in animals, some nuclear RNA is composed of single-copy DNA transcripts covalently linked to sequences transcribed from repetitive DNA. This hnRNA (heterogeneous nuclear RNA) is the precursor to mRNA, and the repetitive portions are transcribed from both, the interspersed repetitive DNA and introns within the genes (e.g., Costantini *et al.*, 1978, 1980; Ryffel *et al.*, 1981; Elder *et al.*, 1981; Balmain *et al.*, 1982; Colman, 1983). As the transcribed repetitive sequences contain those of the *Alu* family and the snRNAs (which occur evidently in all animal and plant species: Blin *et al.*, 1983), various functional significance has been speculated but not proved so far (e.g., RNA processing, or regulation of gene activation in the sense of the model of Britten and Davidson, 1971; see also Davidson and Posakony, 1982; Davidson *et al.*, 1983).

Studies comparing the basic DNA amount (2C value), the proportion of con-

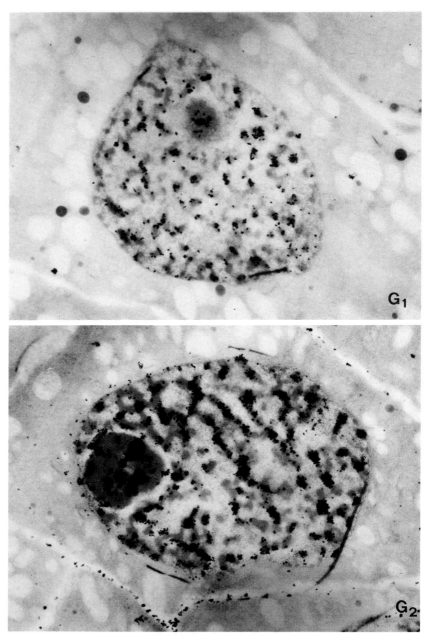

FIG. 9. Electron microscopic autoradiograms of hyacinth root tip nuclei in G_1 and G_2 phase of the cell cycle labeled with [^{3}H]uridine. The label is restricted to the nucleoli and the condensed chromatin. ×6500.

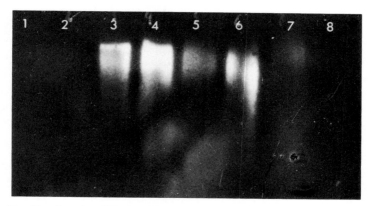

FIG. 10.   Negative print of Northern blot hybridization of gels with smears different preparations of polysomal messenger RNA of *Vicia faba* root tips with radioactive labeled highly repetitive DNA (c_0t 0–0.001) of the same species. Note the heavy signals, which are missing in controls (gel 1; ϕX DNA; gel 2; *Vicia* total DNA; gel 8; *Vicia* ribosomal DNA).

densed chromatin, and the average rate of RNA synthesis per picogram DNA in angiosperm species with different 2C value revealed that there is a negative correlation between DNA amount and condensed chromatin on the one hand, and transcriptional activity on the other hand (Nagl, 1982a,b). Nevertheless, the amount of transcribed DNA is larger in species with high 2C values, although the number of genes is expected to be the same or nearly so. These findings led to the suggestion that repetitive DNA, probably located within condensed chromatin, is transcribed in higher plants. Unpublished investigations employing both electron microscope autoradiography of *Hyacinthus* nuclei after [³H]uridine incorporation and hybridization of repetitive DNA to gels of polysomal RNA of *Vicia* confirmed this suggestion (Figs. 9 and 10). Uridine incorporation has also been found in heterochromatin of grasshoppers (Majumdar *et al.*, 1979), and some of the electron micrographs of *Lilium* microsporophytes shown by Porter *et al.* (1982) display silver grains over dense chromatin, too. This is again in contrast to the findings in mammalian cells, where transcription is always restricted to decondensed, diffuse chromatin.

C. DNA REPLICATION AND CHROMATIN REPLICATION

DNA replication itself in all its complexity is only the one side of chromosome replication in eukaryotes. The other side is the synthesis of chromosomal proteins, mainly histones, and their assembly with DNA, normally called chromatin replication or chromatin biosynthesis. Although the mode of histone distribution in replicating DNA is not yet clear (conservative to one daughter double helix, semiconservative in the form of half chromosomes, or randomly in groups of nucleosomes to the one and the other daughter molecule), the results obtained in animals and in *Vicia* (Yakura and Tanifuji, 1980) indicate the following. During or shortly after DNA is replicated, at least some parts of the DNA double strand containing newly synthesized DNA intermediates are organized in the nucleosome structure. However, the frequency and spacing of nucleosomes in the nascent chromatin are different from those in the nonreplicating chromatin. As the replication proceeds, the shortened spacers begin to elongate rapidly and the normal spacing of nucleosomes and the normal susceptibility to nuclease attack are restored, even when a great part of the newly synthesized DNA strand is still in the form of intermediate molecules before ligation.

At the level of condensed chromatin as visible in the light and electron microscope, the issue is somewhat controversial, although this might be a matter of methods employed and of interpretations rather than of reality. It seems that part of the condensed chromatin undergoes some decondensation during DNA replication, while other parts do not (Nagl, 1977a; reviewed by Nagl, 1977b). In general, it is now (in contrast to earlier conclusions) evident that no large-scale, prolonged decondensation is essential for replication, (Nagl, 1977b; Setterfield *et al.*, 1978).

D. Chromatin Organization and the Cell Cycle

It is well known that chromatin of mitotically active cells occurs in two forms: the decondensed interphase chromatin (except heterochromatin), and the specifically condensed mitotic chromatin (or chromosomes). During the last years evidence accumulated of more difficult changes of chromatin organization throughout the cell cycle. For instance, nuclei in G_1 and G_2 of the mitotic and endomitotic cycle can be distinguished in some animals and plants even at the light microscope level (Altmann, 1966; Nagl, 1968).

Recently, it became possible to digitize and quantitize cell cycle-specific changes in nuclear structure (e.g., Sans and de la Torre, 1979a; Sawicki, 1979; Nicolini, 1979, 1980a; Bibbo *et al.*, 1981). Also at the electron microscope level, computer-aided automated identification and digitation of nuclear images is now possible in plants (Fig. 11; Nagl, 1983a) and animals. However, some uncertainties still exist in the interpretation of some stages with respect to their position in the cell cycle, due to difficulties in ranging them appropriately. In higher plants, this is especially valid for the so-called dispersion stage (or z phase, or *Zerstäubungs-Stadium*). Originally, this stage was described by Heitz (1929) to be the expression of the onset of mitosis, and to be characterized by the decondensation of heterochromatin. Later, it was thought that this phase is necessary in order to separate the chromosomes which were replicated during the S phase (Nagl, 1968). This dispersion stage was also said to be the expression of endomitosis in plants (Geitler, 1953; Tschermak-Woess and Hasitschka, 1953).

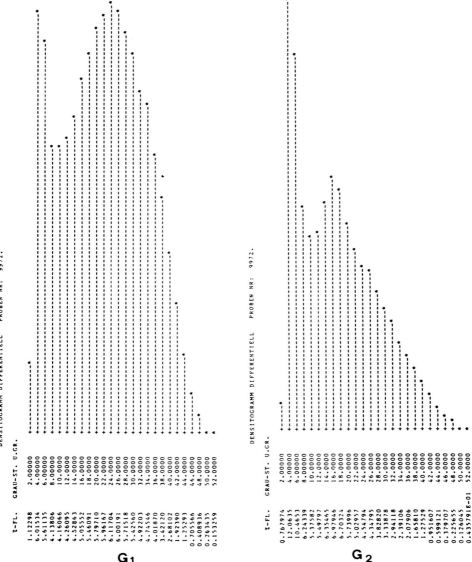

FIG. 11. Computer-drawn densitograms of a Feulgen-stained G_1 nucleus and a G_2 nucleus from a *Lathyrus odoratus* embryo (Leitz T. A. S. image analyzing system); modified from Nagl (1983a).

More recently, doubts have been expressed concerning this interpretation, because [3H]thymidine incorporation was found to occur into the decondensed heterochromatin. Therefore, Barlow (1976) assumed that the disperson stage is identical with the late S period, when the heterochromatin is asynchronously replicating. It may also be that partial decondensation of heterochromatin (i.e., down to the state of condensed euchromatin) takes place twice, during late S period and early prophase (Nagl, 1977a).

In mammals, a continuous decondensation from telophase to late S phase, and a continous condensation from early G_2 to anaphase was postulated by Mazia (1963). Such a cycle of decondensation, replication, condensation, and distribution (segregation) was confirmed by various methods with some deviations in the location of the phase of maximum decondensation (which was, for instance, found at the beginning of the S period instead of its end; see Rao and Hanks, 1980). Kendall *et al.* (1977) and Nicolini (1979, 1980a, 1982) reported, however, on the basis of studies using an automated image analyzer, that there occur two decondensation–condensation events between telophase and the S period. This could, however, also be due to changes in the nuclear volume during G_1 (Rao and Hanks, 1980). For detailed studies of certain stages see Sheinin *et al.* (1980), Dolby *et al.* (1981), and Moser *et al.* (1981). Parameters for the discrimination G_0 and G_1 cells, and G_{01} and G_{02} cells, in animals and plants were given by Nicolini (1980a,b) and Sans and de la Torre (1979b).

The causes for the cell cycle-related changes in chromatin organization can be seen in the nonconcordant increase in chromatin mass and nuclear volume during replication in changes in the nuclear lamina, chromatin conformational changes during the decondensation–condensation cycle, histone H1 phosphorylation, and changes in transcriptional activity (Nagl, 1968; Gurley *et al.*, 1978; Dolby *et al.*, 1981; Jost and Johnson, 1981). The latter aspect is consistent with the fact that many cell cycle-specific gene activities and cell properties are known (reviewed by Nagl, 1976a, 1978).

E. Chromatin Transformation during Carcinogenesis

For reasons of being complete, I would like to draw the reader's attention briefly to alterations of chromatin organization during carcinogenesis. These changes appear to me to be especially important in understanding the close relationship between an altered conformation of chromatin on the one hand, and an altered pattern of gene activities and cell functions on the other hand. Moreover, such changes in chromatin structure are an important aid in the early diagnosis of transformation and of the malignant state. The changes observed are related to two properties of neoplastic tissues. First, transformed cell populations show some progression into S and G_2 phases and exhibit, therefore, the related structures (e.g., Hittelman and Rao, 1978; Nagl and Hiersche, 1983). Second, the switch of gene activity from the tissue-specific pattern to tne neoplastic pattern, as well as chromosomal aberrations, are expressed in a change in the design and amount of condensed chromatin, the nuclear size, and nuclear shape (Sprenger *et al.*, 1973; Nicolini, 1979, 1980a,b; Romen *et al.*, 1979; Hartwig *et al.*, 1980; Leonardson and Levy, 1980; Monaselidze *et al.*, 1980; Anders *et al.*, 1981; Stegner, 1981; Herzog, 1982; Sandritter, 1982; and many others).

It is interesting to note that chromatin changes during carcinogenesis involve all levels of chromatin structure. In this connection it is worthwhile to note that nonintegrated viruses are apparently always organized in a nucleosomal manner, at least in part (Griffith, 1975; Chambon, 1977; Gariglio *et al.*, 1979). Unfortunately, nothing is known about the acquisition of transposon DNA from bacterial plasmids in *Agrobacterium tumefaciens*.

VI. Models of Gene Regulation via Chromatin Structure

The regulation of cell-specific gene activity and hence of differentiation and morphogenesis is as poorly understood as the changes in gene regulation during evolution (see Paterson, 1981; Dover, 1982; Stebbins, 1982; Vrba, 1982). Recently, the role of changes in chromatin structure for gene regulation has been accepted by many cell biologists, but the question of how they are controlled themselves is still a field of speculation. Some suggested models will very briefly be shown here (they all are mainly based on animal systems).

The role of nonhistone chromosomal proteins in decondensing specific chromatin regions has been mentioned already in Section V,A. As these proteins represent gene products themselves, they cannot be taken as the basic regulators, because their own genes must be activated before, and so on *ad infinitum*. The same is valid for any other regulatory protein such as a histone phosphokinase. A similar problem is found with hormones. They are synthesized only in specific cells, and—if transferred through the organism—they can only affect cells which provide the appropriate receptor, i.e., which are already predifferentiated. And in the case of evolution, we will find similar problems. The genes changed only very little in number and function, in any case much too little to make them responsible for the large morphological and physiological changes (see the discussion by Nagl, 1979b, 1983c). Therefore, recent models take into account that there must exist some physical and physicochemical properties of living matter, which are the basis of regulation and which do not require further control, because they are part of self-organizing systems.

A. The Concept of Conformational DNA

Let us consider a few relevant facts. The genomes of plants and animals underwent prominent changes during evolution, thus specifying evolutionary novelty and organismic complexity. It is, however, now evident that the nuclear

DNA content varies among organisms over a range of about $1:10^5$ without a similar change in the number of genes. In higher organisms only 1.0 to 0.1% or less of the DNA represent genes (for references see Nagl, 1982a,b, 1983c), while the mass of the DNA envoled by amplification and diversification of noncoding sequences (e.g., Flavell et al., 1979). Moreover, the genes code for very similar proteins in all higher organisms, in spite of many time-dependent mutations. It was, therefore, suggested that this DNA may be involved in the speciation process, and not so much in adaptation (Nagl, 1977, 1978; Grant, 1981). Recently it was shown that the species-specific chromatin organization in plants (as visible as nuclear structure and ultrastructure, respectively) is dependent on the 2C DNA content and the proportion of repetitive DNA fractions (Nagl, 1982a; Nagl et al., 1983b). This is consistent with the view that the non-protein-coding DNA is involved in the control of chromatin gross organization. (I suggested therefore, the term "conformational DNA"; Nagl 1983b,c.) It seems very unlikely that the huge amount of nongene DNA is just selfish, or parasitic, or junk, as suggested by some authors (e.g., Doolittle and Sapienza, 1980). But it seems to be very likely that there exist some basic physical factors which underlie the diversification of biomatter (reviewed by Popp, 1979; Fröhlich, 1980, 1983; Haken, 1980a; Nagl, 1983b,c).

It is well established that DNA carries the information required for its actual conformation in three dimensions by the sequence of its elements (e.g., Rhodes and Klug, 1981; Sasisekharan et al., 1981). The structure can be predicted by applying topological and thermodynamic rules for finding the energetically most favorable structure for a given sequence, whereby the solvent plays an important role (e.g., Albergo and Turner, 1981).

It is also known that higher order structures of DNA, i.e., architecture of the chromatin in the nucleus, is influenced by the nucleotide sequence organization (Flavell, 1980; Nagl, 1982a,b). Chromatin organization must, however, have a profound effect on the way genes work in cells (Nicolini, 1979, 1980a; Smith, 1981; Zuckerkandl, 1981). When a cell differentiates, that is, becomes specialized, a particular suite of genes pertinent to the cell's future activity is put into a state of readiness for transcription. This involves a different conformation of chromatin preceding transcription; the genes and a large area around them are now sensitive to DNase I. These conformational changes are not always visible at the light and electron microscopic levels in plants (perhaps because less than 0.1% of the chromatin is in the active state), but in mammals they are clearly visible. Also misdifferentiation in the case of malignant events is reflected by conformational changes of chromatin.

Experimental evidence for site-specific responses of chromatin domains to rather unspecific chemical and physical factors comes from studies in salivary gland chromosomes. Kroeger and co-workers (Kroeger and Müller, 1973; Lezzi, 1967; Robert et al., 1978) reported that certain concentrations and combinations of ions selectively stimulated one or the other puffing pattern in isolated polytene nuclei. These results can only be understood in terms of different conformations of chromatin in the single bands of the polytene chromosomes (caused by the nucleotide sequence and the specific binding of nonhistone proteins as polyanions). According to their conformation the bands respond, or do not, to a change of the physicochemical environment according to thermodynamic and electrodynamic rules. Well-known examples of specific gene activation by an unspecific physical factor are also the syntheses of heat shock proteins. This kind of control is not possible on genes directly, because their base sequence is dictated by the genetic code. A similar situation can be expected in nonpolytenic cells. This ideas lead to another biophysical theory.

B. THE POLYELECTROLYTE THEORY

According to polyelectrolyte theory (Manning, 1978; Nicolini, 1979, 1980a), the ionic milieu is reflected parallelly by changes in chromosomal proteins and bound ions (monovalent and bivalent). DNA conformation is naturally dependent on the extent of electrostatic repulsion of the negatively charged phosphate groups. Calculated changes in free energy for small angle bending coupled to DNA persistance length measurements at various ionic swrengths, indicate that there is an entropy-drived negative change in free energy, $\Delta G \_ 77$ cal/mol/base pairs, which would lead to spontaneous bending of DNA if the phosphate were neutralized. Thus, even a small change in univalent ion concentration may cause abrupt transitions in higher order chromatin structure (for the role of the pH see Labhart et al., 1981).

C. EXCITED STATES OF DNA AND ELECTROMAGNETIC COUPLINGS

Considering organisms as dissipative structures (Prigogine, 1955), the thermodynamics of irreversible processes in open systems far away from equilibrium have clearly shown that negentropy transitions are, under appropriate conditions, physically necessary and an important factor in the evolution of life and net increase in complexity ("order through fluctuations": Schrödinger, 1944; Prigogine, 1976; Lamprecht and Zotin, 1978; Gladyshev, 1982). With respect to Fröhlich's (1968, 1980) fundamental papers on energy storage and long-range coherence in biological systems and the basic work of Prigogine (1955), Nicolis and Prigogine (1977), and Haken (1980a,b) we feel that the aspect of coherence may play an important, if not the most essential role in order to describe and understand biological phenomena. This stimulated investigations on characteristics of biological photon emission (reviewed by Slawinska and Slawinski, 1983) with respect to its possible coherence. It was shown that (1) there are some indications of at least partial coherence of ultraweak radiation in biological systems, (2) these features can be described in terms of coherent DNA–exciplex formation, and (3) a variety of biological phenomena can be understood within the framework of this model (Popp, 1979; Popp *et al.*, 1981, 1984a,b; Li *et al.*, 1983; Nagl and Popp, 1983; Popp and Nagl, 1983).

An important aspect of macromolecular conformation and energetics is their electronic characteristics (Lavery and Pullman, 1982). Interplanar interactions of nucleic bases, the so-called electronic stacking interactions, make an important contribution to the stability of the structure and functional properties of DNA. Stacking interactions are characterized by negative values of changes of entropy (ΔS), enthalpy (ΔH), and partial volume (ΔV) (Maevsky *et al.*, 1980). As in other systems, therefore, thermodynamic (and quantum mechanical) constraints may well be envisaged as the basic factors of evolution of genome organization and hence of diversification of organisms.

Great biological significance can perhaps be seen in the fact that DNA could act as a photon trap and photon emittor (Popp, 1979; Popp *et al.*, 1981; Rattemeyer *et al.*, 1981). In the short time between the absorption of a photon (quantum) by one of the bases of DNA and the final storage or emission, many electronic rearrangements can take place, various excited states of different multiplicities may be populated, energy may be transferred between different regions of the DNA, and some hydrogen bonds and even covalent bonds may be formed while others may be broken (see Eisinger and Shulman, 1968; Morgan and Daniels, 1980). The bases have much stronger interactions in the excited states and these interactions again can profoundly affect the excited states.

It is now well established that polynucleotides emit photons at room temperature, which come from excimers. Excimers (excited dimers) are complexes of two molecules, with an attractive binding potency, because at least one molecule is excited by photons. The emission occurs evidently in part in a coherent way (as in a laser), what can be seen as an elementary process of matter in nonequilibrium states (Li, 1981; Popp *et al.*, 1981; see also Fröhlich, 1980). This emission can lead to structural information on the polynucleotide in solution, but it is also important since it is known that excimer formation represents an excitation energy trap (Vigny and Duquesne, 1976; see also Englander *et al.*, 1980). The driving force of excitation of DNA and increase in molecular information complexity may be seen in a Bose–Einstein condensation (Wu and Austin, 1978; Mishra *et al.*, 1979). The capacity of photon storage and the intensity of photon emission in theory strongly depend on the degree of chromatin condensation and hence DNA conformation. Therefore, conformational DNA may also have an important function as a photon trap and photon emittor, and by this way control many metabolic and cellular events via electromagnetic, cooperative phenomena. These examples show that the individual parts (sequences) of the system (DNA) cooperate in producing macroscopic spatial, temporal, and functional structures, i.e., we deal with a synergetic system (Haken, 1980a,b). Fig-

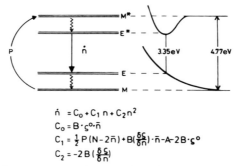

$$\acute{n} = C_0 + C_1 n + C_2 n^2$$
$$C_0 = B \cdot \varsigma^0 \cdot \bar{n}$$
$$C_1 = \tfrac{1}{2} P(N - 2\bar{n}) + B(\tfrac{\delta \varsigma}{\delta n}) \cdot \bar{n} - A - 2B \cdot \varsigma^0$$
$$C_2 = -2B(\tfrac{\delta \varsigma}{\delta n})$$

FIG. 12. Energy diagram corresponding to a four-niveau laser system as a model of photon storage and emission by, and from, DNA respectively (*N*; number of base pairs; *ñl*; total number of excimers); modified from Li *et al.* (1983).

ure 12 shows the energy diagram of DNA corresponding to a four-niveau system. k_4 represents the radiationless transition rate from level 4 to level 3, and k_2 is that from level 2 to 1. A, B are the Eistein's coefficients for spontaneous and induced transitions, respectively. n_i is the occupation number of energy level i, ρ represents the energy density of the radiation field due to transitions between level 3 and 2. Details and calculations of the model are given in Li *et al.* (1983).

VII. Conclusions

The conclusions which can be drawn from the findings on chromatin organization are the following.

There exists a complex feedback control system between genome structure and chromatin structure. The DNA amount, the proportion of repetitive DNA, the arrangement of sequences etc. determine the species-specific nuclear structure (percentage and pattern of condensed chromatin) in higher plants (Nagl *et al.*, 1983b). As shown in animal systems, DNA sequences located upstream of genes have a role in determining the chromatin structure of the surrounding region (DNase I-hypersensitive sites; Elgin, 1982). This chromatin structure, on the countereffect, enables gene activation. Changes in DNA conformation caused by ions (e.g., a B to Z transition—the Z conformation has been found by indirect immunoflourescence in eukaryotic chromosomes: Hamada *et al.*, 1982; Lemeunier *et al.*, 1982; Viegas-Péquignot *et al.*, 1982) leads to a different protein binding capacity (e.g., Nordheim *et al.*, 1982) and hence structural changes of chromatin. Transposition of DNA sequences by transposable elements or by chromosomal aberrations leads to an altered arrangement of chromatin domains (and hence genes) within the nucleus. This can lead to differentiation or carcinogenesis, because of direct influences of chromatin conformation on the accessibility of the transposed gene, or by long-range interactions within the nucleus (like position effects in short distances; Macieira-Coelho, 1980; Strong, 1980; Hilliker *et al.*, 1980; Weinstein, 1981).

Therefore, gene activity is evidently controlled via chromatin conformation, which itself is the result of genome organization plus the electrostatic environment. The local changes of chromatin conformation during development occur according to thermodynamic and electrodynamic, lastly quantum mechanical rules, and are dependent on the position of a cell within various physical and chemical gradients (Nagl, 1983c). The altered chromatin structure then allows the action of finer tuning mechanisms of gene regulation as suggested by the Jacob–Monod model or the Britten–Davidson model. In addition, the conformation of chromatin and DNA may lead to certain electronic states of these molecules with different levels of excitation (excimer, exciplex formation, and decay) and (perhaps coherent) photon emission with feedback effects on structure and function of the nucleus and the cell. In general, biologists will have to reformulate many problems in terms of collective and nonlinear phenomena (Rowlands, 1983) and in terms of physical determinism (Lima-de-Faria, 1983).

ACKNOWLEDGMENTS

I am grateful to Professor Dieter Schweizer, Vienna, for the plasmid gift, to Miss Silvia Kühner for careful technical assistance, and to Misters Thomas Weber and Hans-Peter Schmitt for providing unpublished figures. I thank Miss Elisabeth Miesel for typing the manuscript.

REFERENCES

Albergo, D. D., and Turner, D. H. (1981). *J. Cell Biol.* **90,** 181–186.

Altmann, H. -W. (1966). *Verh. Dtsch. Ges. Pathol.* **50,** 15–51.

Anders, F., Chatterjee, K., Schwab, M., Scholl, E., and Anders, A. (1981). *Am. Zool.* **21,** 535–548.

Antoine, J. L., Aurias, A., and Dutrillaux, B. (1982). *Ann. Genet.* **25,** 226–228.

Avivi, L., Feldman, M., and Brown, M. (1982a). *Chromosoma* **86,** 1–16.

Avivi, L., Feldman, M., and Brown, M. (1982b). *Chromosoma* **86,** 17–26.

Bak, A. L., and Zeuthen, J. (1977). *Cold Spring Harbor Symp. Quant. Biol* **42,** 367–377.

Balmain, A., Frew, L., Cole, G., Krumlauf, R., Ritchie, A., and Birnie, G. D. (1982). *J. Mol. Biol.* **160,** 163–179.

Barlow, P. W. (1976). *Protoplasma* **90,** 381–392.

Barlow, P. W. (1977). *Ann. Sci. Nat. Bot. Biol. Veg.* **18,** 193–206.

Barlow, P. W. (1982). *In* "Growth Regulators in Plant Senescence" (M. B. Jackson, B. Grout, and I. A. Mackenzie, eds.), pp. 27–45. Brit. Plant Growth Regul. Grp. Monogr. 8, London.

Barlow, P. W., and Barlow, S. (1980). *Annu. Rep. Aust. Natl. Univ., Biol. Sci.* Abstr. No. 21.

Barlow, P. W., and Hines, E. (1980). *Annu. Rep. A. R. C. Letcombe Lab. Wantage* 70.

Barlow, P. W., Rost, T. L., and Gunning, B. E. S. (1982). *Protoplasma* **112**, 205–216.

Bennett, M. D. (1973). *Brookhaven Symp. Biol.* **25**, 344–366.

Bennett, M. D. (1982). *In* "Genome Evolution" (G. A. Dover and R. B. Flavell, eds.), pp. 239–261. Academic Press, New York.

Berlin, J., Knobloch, K. -H., Höfle, G., and Witte, L. (1982). *J. Nat. Prod.* **45**, 83–87.

Biard-Roche, J., Gorka, C., and Lawrence, J. -J. (1982). *EMBO J.* **1**, 1487–1492.

Bibbo, M., Bartels, P. H., Sychra, J. J., and Wied, G. L. (1981). *Acta Cytol.* **25**, 23–28.

Blin, N., Weber, Th., and Alonso, A. (1983). *Nucleic Acids Res.* **11**, 1375–1388.

Brasch, K., and Sinclair, G. D. (1978). *Virchows Arch. B* **27**, 193–204.

Britten, R. J., and Davidson, E. H. (1971). *Q. Rev. Biol.* **46**, 111–138.

Brown, I. R. (1980). *Dev. Biol.* **80**, 248–252.

Brown, I. R. (1982). *Biochim. Biophys. Acta* **698**, 307–309.

Burkhanova, E., Staynov, D. Z., Koleva, St., and Tsanev, R. (1975). *Cell Differ.* **4**, 201–207.

Capesius, I., and Nagl, W. (1978). *Plant Syst. Evol.* **129**, 143–166.

Cartwright, I. L., Abmayr, S. M., Fleischmann, G., Lowenhaupt, K., Elgin, S. C. R., Keene, M. A., and Howard, G. C. (1982). *CRC Crit. Rev. Biochem.* **13**, 1–86.

Cavalier-Smith, T. (1982). *Annu. Rev. Biophys. Bioeng.* **11**, 273–302.

Chambon, P. (1977). *Cold Spring Harbor Symp. Quant. Biol.* **42**, 1211–1236.

Colman, A. (1983). *TIBS* **8**, 37.

Comings, D. E. (1978). *Annu. Rev. Genet.* **12**, 25–46.

Costantini, F. D., Scheller, R. H., Britten, R. J., and Davidson, E. H. (1978). *Cell* **15**, 173–187.

Costantini, F. D., Britten, R. J., and Davidson, E. H. (1980). *Nature (London)* **287**, 111–117.

Cremer, T., Cremer, C., Baumann, H., Luedtke, E. K., Sperling, K., Teuber, V., and Zorn, C. (1982). *Hum. Genet.* **60**, 46–56.

Davidson, E. H., and Britten, R. J. (1979). *Science* **204**, 1052–1059.

Davidson, E. H., and Posakony, J. W. (1982). *Nature (London)* **297**, 633–635.

Davidson, E. H., Jacobs, H. T., and Britten, R. J. (1983). *Nature (London)* **301**, 468–470.

Derenzini, M., Pession-Brizzi, A., Betts-Eusebi, C., and Novello, F. (1981). *J. Ultrastruct. Res.* **75**, 229–242.

Derenzini, M., Hernandez-Verdun, D., and Bouteille, M. (1982). *Exp. Cell Res.* **141**, 463–469.

Dhillon, S. S., and Miksche, J. P. (1981). *Physiol. Plant.* **51**, 291–298.

Dhillon, S. S., and Miksche, J. P. (1982). *Am. J. Bot.* **69**, 219–226.

Dhillon, S. A., and Miksche, J. P. (1983). *Histochem. J.* **15**, 21–38.

Dolby, T. W., Belmont, A., Borun, T. W., and Nicolini, C. (1981). *J. Cell Biol.* **89**, 78–85.

Doolittle, W. F., and Sapienza, C. (1980). *Nature (London)* **284**, 601–603.

Dover, G. (1982). *Nature (London)* **299**, 111–117.

Drumm, N., and Nagl, W. (1982). *Mech. Ageing Dev.* **18**, 103–110.

DuPraw, E. J. (1970). "DNA and Chromosomes". Holt, New York.

Eisinger, J., and Shulman, R. G. (1968). *Science* **161**, 1311–1319.

Elder, J. T., Pan, J., Duncan, C. H., and Weissman, S. M. (1981). *Nucleic Acids Res.* **9**, 1171–1189.

Elgin, S. C. H. (1981). *Cell* **27**, 413–415.

Elgin, S. C. H. (1982). *Nature (London)* **300**, 402–403.

Englander, S. W., Kallenbach, N. R., Heeger, A. J., Krumhansl, J. A., and Litwin, S. (1980). *Proc. Natl. Acad. Sci. U.S.A.* **77**, 7222–7226.

Erenpreisa, J. (1982). *Indian J. Exp. Biol.* **20**, 507–512.

Erenpreiss, J. G., Zirne, R. A., and Demidenko, O. Je. (1981). *Folia Histochem. Cytochem.* **19**, 217–228.

Evans, K. J., and Filion, W. G. (1982). *Can. J. Genet. Cytol.* **24**, 583–591.

Felsenfeld, G. (1978). *Nature (London)* **271**, 115–122.

Finch, J. T., and Klug, A. (1976). *Proc. Natl. Acad. Sci. U.S.A.* **73,** 1897–1901.
Flavell, R. (1980). *Annu. Rev. Plant Physiol.* **31,** 569–596.
Flavell, R. B., Bedbrook, J., Jones, J., O'Dell, M., and Gerlach, W. L. (1979). *Proc. John Innes Symp.* 15–30.
Franke, W. W., Scheer, U., Trendelenburg, M. F., Spring, H., and Zentgraf, H. (1976). *Cytobiologie* **13,** 401–434.
Frisch, B., and Nagl, W. (1979). *Plant Syst. Evol.* **131,** 261–276.
Fröhlich, H. (1968). *Int. J. Quantum Chem.* **2,** 641–656.
Fröhlich, H. (1980). *Adv. Electron. Electron Phys.* **53,** 85–152.
Fröhlich, H., ed. (1983). "Coherent Excitations in Biological Systems." Springer-Verlag, Berlin and New York.
Gariglio, P., Llopis, R., Oudet, P., and Chambon, P. (1979). *J. Mol. Biol.* **131,** 75–105.
Gärtner, P. -J., and Nagl, W. (1980). *Planta* **149,** 341–349.
Geitler, L. (1953). "Endomitose und endomitotische Polyploidisierung." Springer-Verlag, Berlin and New York.
Gerlach, W. L., and Bedbrook, J. R. (1979). *Nucleic Acids Res.* **7,** 1869–1885.
Gigot, C., Philipps, G., Nicolaieff, H., and Hirth, L. (1976). *Nucleic Acids Res.* **3,** 2315–2329.
Gladyshev, G. P. (1982). *J. Theor. Biol.* **94,** 225–239.
Goldowitz, D., and Mullen, R. J. (1982). *Dev. Biol.* **89,** 261–267.
Grant (1981). *Biol. Zbl.* **100,** 473–482.
Gray, P., and Nagl, W. (1981). *Nucleus* **24,** 1–3.
Greimers, R., and Deltour, R. (1981). *Eur. J. Cell Biol.* **23,** 303–311.
Griffith, J. D. (1975). *Science* **187,** 1202–1203.
Guerri, J., Culianez, F., Primo-Millo, E., and Primo-Yúfera, E (1982). *Planta* **155,** 273–280.
Gurley, L. R., D'Anna, J. A., Barham, S. S., Deaven, L. L., and Tobey, R. A. (1978). *Eur. J. Biochem.* **84,** 1–15.
Hager, H. D., Schroeder-Kurth, T. M., and Vogel, F. (1982). *Hum. Genet.* **61,** 342–356.
Haken, H. (1980a). *Naturwissenschaften* **67,** 121–128.
Haken, H. ed. (1980b). "Dynamics of Synergetic Systems." Springer-Verlag, Berlin and New York.
Hamada, H., Petrino, M. G., and Kakunaga, T. (1982). *Proc. Natl. Acad. Sci. U.S.A.* **79,** 6465–6469.
Hartwig, M. (1982). *Biochim. Biophys. Acta* **698,** 214–217.
Hartwig, M., Matthes, E., and Arnold, W. (1980). *Stud. Biophys.* **79,** 17–18.
Havelange, A. (1980). *Am. J. Bot.* **67,** 1171–1178.
Havelange, A., and Bernier, G. (1974). *J. Cell Sci.* **15,** 633–644.
Heitz, E. (1929). *Ber. Dtsch. Bot. Ges.* **47,** 274–284.
Herzog, R. E. (1982). *Arch. Gynecol.* **231** 91–98.
Hesemann, C. U., and Schröder, G. (1982). *Theor. Appl. Genet.* **62,** 325–328.
Hilliker, A. J., Appels, R., and Schalet, A. (1980). *Cell* **21,** 607–619.
Hirahara, S. (1980). *J. Sci. Hiroschima Univ., Ser. B* **17,** 9–49.
Hittelman, W. N., and Rao, P. N. (1978). *J. Cell. Physiol.* **95,** 333–337.
Hsu, T. C. (1962). *Exp. Cell Res.* **27,** 332–334.
Jacqmard, A., Miksche, J. P., and Bernier, G. (1972). *Am. J. Bot.* **59,** 714–721.
Jacqmard, A., Kettmann, R., Pryke, J. A., Thiry, M., and Sachs, R. M. (1981). *Ann. Bot.* **47,** 415–417.
Jeanny, J. -C., and Gontcharoff, M. (1978). *Wilhelm Roux's Arch.* **184,** 195–211.
Jeanny, J. C., and Gontcharoff, M. (1981). *J. Histochem. Cytochem.* **29,** 1281–1288.
Jost, E., and Johnson, R. T. (1981). *J. Cell Sci.* **47,** 25–53.
Kendall, F., Swenson, R., Borun, T., Rowinski, J., and Nicolini, C. (1977). *Science* **196,** 1106–1109.

Kiefer, G., Kiefer, R., Moore, G. W., Salm, R., and Sandritter, W. (1974). *J. Histochem. Cytochem.* **22**, 569–576.

Kornberg, R. D. (1977). *Annu. Rev. Biochem.* **46**, 931–954.

Kroeger, H., and Müller, G. (1973). *Exp. Cell Res.* **82**, 89–94.

Labhart, P., and Koller, T. (1982). *Cell* **28**, 279–292.

Labhart, P., Thoma, F., and Koller, T. (1981). *Eur. J. Cell Biol.* **25**, 19–27.

Lafontaine, J. G. (1974). *In* "The Cell Nucleus" (H. Busch, ed.), Vol. I, pp. 149–185. Academic Press, New York.

Lai, V., and Srivastava, L. M. (1976). *Cytobiologie* **12**, 220–243.

Lamprecht, I., and Zotin, A. I. (1978). "Thermodynamics of Biological Processes." De Gruyter, Berlin.

Lavery, R., and Pullman, B. (1982). *Nucleic Acids Res.* **10**, 4383–4395.

Leber, B., and Hemleben, V. (1979). *Nucleic Acids Res.* **7**, 1263–1281.

Lemeunier, F., Derbin, C., Malfoy, B., Leng, M., and Taillandier, E. (1982). *Exp. Cell Res.* **141**, 508–513.

Leonardson, K. E., and Levy, S. B. (1980). *Nucleic Acids Res.* **8**, 5317–5331.

Leweke, B., and Hemleben, V. (1982). *In* "The Cell Nucleus" (H. Busch, ed.), Vol. XI, pp. 225–253. Academic Press, New York.

Lezzi, M. (1967). *Chromosoma* **21**, 109–122.

Li, K. (1981). *Laser Elektroopt.* **3**, 32–35.

Li, K. H., Popp, F., Nagl, W., and Klima, H. (1983). *In* "Coherent Excitations in Biological Systems" (H. Fröhlich, and F. Kremer, eds.), pp. 141–146. Springer-Verlag, Berlin and New York.

Liberati-Langenbuch, J., Wilhelm, M. L., Gigot, C., and Wilhelm F. X. (1980). *Biochem, Biophys. Res. Commun.* **94**, 1161–1168.

Lima-de-Faria, A. (1983). "Molecular Evolution and Organization of the Chromosome." Elsevier, Amsterdam.

Lohr, D. (1983). *Biochemistry* **22**, 927–934.

Lutz, C., and Nagl, W. (1980). *Planta* **149**, 408–410.

McGhee, J. D., and Felsenfeld, G. (1980). *Annu. Rev. Biochem.* **49**, 1115–1156.

Macieira-Coelho, A. (1980). *Gerontology* **26**, 276–282.

Maevsky, A. A., Sarvazyan, A. P., and Hemmes, P. (1980). *Stud. Biophys.* **79**, 79–80.

Maio, J. J., Brown, F. L., and Musich, P. R. (1977). *J. Mol. Biol.* **117**, 637–655.

Majumdar, K., Chatterjee, C., and Ray-Chaudhuri, S. P. (1979). *Nucleus* **22**, 149–152.

Manning, G. (1978). *Q. Rev. Biophys.* **11**, 179–246.

Marsden, M. P. F., and Laemmli, U. K. (1979). *Cell* **17**, 849–858.

Mazia, D. (1963). *J. Cell. Comp. Physiol. Suppl. 1*, **62**, 123–140.

Mirzabekov, A. D. (1980). *Q. Rev. Biophys.* **13**, 255–295.

Mishra, R. K., Bhaumik, K., Mathur, S. C., and Mitra, S. (1979). *Int. J. Quantum Chem.* **16**, 691–706.

Mitchell, J. P., and van der Ploeg, M. (1982). *Histochemistry* **75**, 327–340.

Monaselidze, J. R., Chanchalshvili, Z. I., Chitadze, G. V., Majagaladze, G. V., and Mgeladze, G. N. (1980). *Stud. Biophys.* **81**, 173–174.

Moreno, M. L., Sogo, J. M., and de la Torre, C. (1978). *New Phytol.* **81**, 681–683.

Morgan, J. P., and Daniels, M. (1980). *Photochem. Photobiol.* **31**, 207–213.

Moser, G. C., Fallon, R. J., and Meiss, H. K. (1981). *J. Cell. Physiol.* **106**, 293–301.

Müller, U., and Nagl, W. (1983). *Experientia* **39**, 406–407.

Nagl, W. (1968). *Österr. Bot. Z.* **115**, 322–353.

Nagl, W. (1972). *Cytobios* **5**, 145–154.

Nagl, W. (1974). *Z. Pflanzenphysiol.* **73**, 1–44.

Nagl, W. (1976a). "Zellkern und Zellzyklen." Ulmer, Stuttgart.

Nagl, W. (1976b). *Ber. Dtsch. Bot. Ges.* **89**, 301–311.

Nagl, W. (1977a). *Protoplasma* **91**, 389–407.

Nagl, W. (1977b). *In* "Mechanisms and Control of Cell Division" (T. L. Rost and E. M. Gifford, Jr., eds.), pp. 147–193. Dowden, Stroudsburg, Pennsylvania.

Nagl, W. (1978). "Endopolyploidy and Polyteny in Differentiation and Evolution." North-Holland Publ., Amsterdam.

Nagl, W. (1979a). *Protoplasma* **100**, 53–71.

Nagl, W. (1979b). *Plant Syst. Evol., Suppl.* **2**, 247–260.

Nagl, W. (1980). *Microsc. Acta, Suppl.* **4**, 19–25.

Nagl, W. (1981). *Int. Rev. Cytol.* **73**, 21–53.

Nagl, W. (1982a). *In* "Cell Growth" (C. Nicolini, ed.), pp. 171–218. Plenum, New York.

Nagl, W. (1982b). *Encycl. Plant Physiol. New Ser.* **14 B**, 1–45.

Nagl, W. (1983a). *J. Interdiscip. Cycle Res.* **14**, 53–61.

Nagl, W. (1983b). *Proc. Sec. Kew Chrom. Conf.*, **55–61**.

Nagl, W. (1983c). *Biol. Zentralbl.* **102**, 257–269.

Nagl, W., and Bachmann, K. (1980).*Theor. Appl. Genet.* **57**, 107–111.

Nagl, W., and Hiersche, H. D. (1983). *J. Cancer Res. Clin. Oncol.* **105**, A10.

Nagl, W., and Popp. F. A. (1983). *Cytobios* **37**, 45–62.

Nagl, W., Cabirol, H., Lahr, C., Greulach, H., and Ohliger, H. M. (1983a). *Protoplasma* **115**, 59–64.

Nagl, W., Jeanjour, M., Kling, H., Kühner, S., Michels, I., Müller, T., and Stein, B. (1983b). *Biol. Zentralbl.* **102**, 129–148.

Nagl, W. Ribicki, R., Mertler, H.-O., Hezel, U., Jacobi, R., and Bachmann, E. (1984). *Protoplasma* **122**, 138–144.

Nicklisch, A., Strauss, M., and Wersuhn, G. (1976). *Biochem. Physiol. Pflanzen* **169**, 105–119.

Nicolini, C. (1979). *In* "Chromatin Structure and Function" (C. A. Nicolini, ed.), pp. 613–666. Plenum, New York.

Nicolini, C. (1980a). *Cell Biophys.* **2**, 271–290.

Nicolini, C. (1980b). *J. Submicrosc. Cytol.* **12**, 475–505.

Nicolini, C. (1982). *In* "Cell Growth" (C. Nicolini, ed.), pp. 381–434. Plenum, New York.

Nicolini, C., Trefiletti, V., Cavazza, B., Cuniberti, C., Patrone, E., Carlo, P., and Brambilla, G. (1983). *Science* **219**, 176–178.

Nicolis, G., and Prigogine, I. (1977). "Self-Organization in Nonequilibrium Systems." Wiley, New York.

Nordheim, A., Tesser, P., Azorin, F., HaKwon, Y., Möller, A., and Rich, A. (1982). *Proc. Natl. Acad. Sci. U.S.A.* **79**, 7729–7733.

Oshima, R., Curiel, D., and Linney, E. (1980). *J. Supramol. Struct.* **14**, 85–96.

Paiement, J., and Bendayan, M. (1982). *J. Ultrastruct. Res.* **81**, 145–157.

Parslow, T. G., and Granner, D. K. (1982). *Nature (London)* **299**, 449–451.

Paterson, H. E. H. (1981). *S. Afr. J. Sci.* **77**, 113–119.

Philipps, G., and Gigot, C. (1977). *Nucleic Acids Res.* **4**, 3617–3626.

Pierard, D., Jacqmard, A., Bernier, G., and Salmon, J. (1980). *Planta* **150**, 397–405.

Popp, F. A., ed. (1979). "Electromagnetic Bio-Information." Urban & Schwarzenberg, Munich.

Popp, F. A., and Nagl, W. (1983). *Cytobios* **37**, 71–83.

Popp, F. A., Ruth, B., Bahr, W., Böhm, J., Grass, P., Grolig, G., Rattemeyer, M., Schmidt, H. G., and Wulle, P. (1981). *Collect. Phenom.* **3**, 187–214.

Popp, F. A., Nagl, W., Li, K. H., Scholz, W., Weingärtner, O., and Wolf, R. (1984a). *Cell Biophys.* **6**, 33–52.

Popp, F. A., Li, K. H., and Nagl, W. (1984b). *Z. Pflanzenphysiol.* **114**, 1–13.

Porter, E. K., Bird, J. M., and Dickinson, H. G. (1982). *J. Cell Sci.* **57**, 229–246.

Prigogine, I. (1955). "Thermodynamics of Irreversible Processes." Wiley, New York.

Prigogine, I. (1976). *In* "Self-Organization and Social Systems" (E. Jantsch and C. Waddington, eds.), pp. 93–133. Addison-Wesley, Reading, Massachusetts.

Puvion, E., and Viron, A. (1981). *J. Ultrastruct. Res.* **74,** 351–360.

Rao, P. N., and Hanks, S. K. (1980). *Cell Biophys.* **2,** 327–337.

Rattemeyer, M., Popp, F. A., and Nagl, W. (1981). *Naturwiss.* **68,** 572–573.

Rhodes, D., and Klug, A. (1981). *Nature (London)* **292,** 378–380.

Rindt, K. P., and Nover, L. (1980). *Biol. Zbl.* **99,** 641–673.

Robert, M., Bosshard, N., Müller, G., Kühne, W., and Falkenberg, W. (1978). *Verh. Dtsch. Zool. Ges.* **238.**

Rolands, S. (1983). *J. Biol. Phys.* **11,** 117–122.

Romen, W., Roter, A., and Aus, H. M. (1979). *In* "The Automation of Cancer Cytology and Cell Image Analysis" (N. J. Pressman and G. L. Wied, eds.), pp. 47–50. Tutor. of Cytol., Chicago.

Ryffel, G. U., Muellener, D. B., Wyler, T., Wahli, W., and Weber, R. (1981). *Nature (London)* **291,** 429–431.

Sandritter, W. (1982). *Acta Histochem. Suppl.* **26,** 15–14.

Sans, J., and de la Torre, C. (1979a). *Cell Biol. Int. Rep.* **3,** 227–235.

Sans, J., and de la Torre, C. (1979b). *Eur. J. Cell Biol.* **19,** 294–298.

Sasisekharan, V., Gupta, G., and Bansal, M. (1981). *Int. J. Biol. Macromol.* **3,** 2–8.

Sawicki, W. (1979). *In* "Chromatin Structure and Function" (C. A. Nicolini, ed.), pp. 667–681. Plenum, New York.

Scheer, V. (1978). *Cell* **13,** 535–549.

Scheer, V., and Zentgraf, H. (1978). *Chromosoma* **69,** 243–254.

Schmalenberger, B. (1980). *Protoplasma* **103,** 377–391.

Schmalenberger, B., and Nagl, W. (1979). *Plant Syst. Evol Suppl.* **2,** 119–125.

Schrödinger, E. (1944). "What Is Life?" Cambridge Univ. Press, London and New York.

Sedat, J., and Manuelidis, L. (1977). *Cold Spring Harbor Symp. Quant. Biol.* **42,** 331–350.

Setterfield, G., Sheinin, R., Dardick, I., Kiss, G., and Dubsky, M. (1978). *J. Cell Biol.* **77,** 246–263.

Sheinin, R., Setterfield, G., Dardick, I., Kiss, G., and Dubsky, M. (1980). *Can J. Biochem.* **58,** 1359–1369.

Slawinska, D., and Slawinski, J. (1983). *Photochem. Photobiol.* **37,** 709–715.

Sledziewski, A., and Young, E. T. (1982). *Proc. Natl. Acad. Sci. U.S.A.* **79,** 253–256.

Small, D., Nelkin, B., and Vogelstein, B. (1982). *Proc. Natl. Acad. Sci. U.S.A.* **79,** 5911–5915.

Smith, S. G. (1945). *J. Hered.* **36,** 194–196.

Smith, G. R.(1981). *Cell* **24,** 599–600.

Sprenger, E., Moore, G. W., Naujoks, H., Schlüter, G., and Sandritter, W. (1973). *Acta Cytol.* **17,** 27–31.

Stebbins, G. L. (1982). *Evolution* **36,** 1109–1118.

Stegner, H. E. (1981). *Curr. Top. Pathol.* **70,** 171–193.

Strong, L. C. (1980). *Cytobios* **28,** 187–208.

Sugita, M., and Sasaki, K. (1982). *Physiol. Plant,* **54,** 41–46.

Tanaka, N. (1981). *Cytologia* **46,** 545–559.

Tanaka, A. (1982). *Mem. Fac. Educ., Ehime Univ., Ser. III* **2,** 11–77.

Thomas, J. O., and Thompson, R. J. (1977). *Cell* **10,** 633–640.

Tsanev, R. (1974). *Abstr. Proc. FEBS Meet., 9th, Budapest.*

Tsanev, R. (1983). *Molec. Biol. Rep.* **9,** 9–17.

Tschermak-Woess, E. (1963). "Strukturtypen der Ruhekerne von Pflanzen und Tieren." Springer-Verlag, Berlin and New York.

Tschermak-Woess, E., and Hasitschka, G. (1953). *Chromosoma* **5,** 574–614.

Viegas-Péquignot, E., Derbin, C., Lemeunier, F., and Taillandier, E. (1982). *Ann. Genet.* **25**, 218–222.

Vigny, P,, and Duquesne, M. (1976). *In* "Excited States of Biological Macromolecules" (J. Birks, ed.), pp. 167–177. Wiley, New York.

Voges, B., and Fellenberg, G. (1976). *Biochem. Physiol. Pflanzen* **169**, 507–510.

Vrba, E. S. (1982). *S. Afr. J. Sci.* **78**, 275–278.

Weinstein, I. B. (1981). *J. Supramol. Struct. Cell. Biochem.* **17**, 99–120.

Weintraub, H., and Groudine, M. (1976). *Science* **193**, 848–856.

Worcel, A. (1977). *Cold Spring Harbor Symp. Quant. Biol.* **42**, 313–324.

Wu, C. (1982). *In* "Gene Regulation," pp. 147–156. Academic Press, New York.

Wu, T. M., and Austin, S. (1978). *J. Theor. Biol.* **71**, 209–214.

Wu, C., Bingham, P. M., Livak, K. J., Holmgren, R., and Elgin, S. C. R. (1979). *Cell* **16**, 797–806.

Yakura, K., and Tanifuji, S. (1980). *Biochim. Biophys. Acta* **609**, 448–455.

Yakura, K., Fukuei, K., and Tanifuji, S. (1978). *Plant Cell Physiol.* **19**, 1381–1390.

Zhurkin, V. B. (1982). *Stud. Biophys.* **87**, 151–152.

Zuckerkandl, E. (1981). *Mol. Biol. Rep.* **7**, 149–158.

Structure of Metaphase Chromosomes of Plants

GY. HADLACZKY

Institute of Genetics, Biological Research Center, Hungarian Academy of Sciences, Szeged, Hungary

I. Introduction

In recent years a respectable amount of information has accumulated concerning the organization of chromatin. It is currently accepted that the basic chromatin fiber of higher eukaryotes is formed by twisting the DNA double helix around the octamers of core histones (i.e., H2a, H2b, H3, and H4). DNA-core histone particles stabilized by H1 histones and connected by spacer DNA give an ~ 100 Å chain of nucleosomes. A further folding of this basic nucleosome fiber produces the 250–300 Å chromatin fiber which is commonly seen in whole mount electron microscopic preparations of chromosomes or nuclei. The supramolecular organization of this 300 Å chromatin fiber which culminates ultimately in the metaphase chromosome is unelucidated. Despite significant progress in this field the factors involved in the process of higher order organization of metaphase chromosomes, the mechanism of chromosome condensation are not yet clear.

Biochemical decomposition of chromosomes (e.g., DNase digestion, protein extraction) combined with light and electron microscopy has provided substantial data on the structural organization of mammalian chromosome (see for example Maio and Schildkraut, 1966, 1967; Rattner *et al.*, 1975; Mace *et al.*, 1977; Paulson and Laemmli, 1977). As a result of extensive studies on animal chromosomes our knowledge has increased about the role of nonhistones in chromosome architecture (Adolph *et al.*, 1977; Jeppesen *et al.*, 1978) and the organization and packing of chromosomal DNA (Mullinger and Johnson, 1979, 1980). Biochemcial and structural studies on protein-depleted animal chromosomes have led us to the idea that the "structural" nonhistones appear as discrete particles rather than a continuous protein network or core (i.e., scaffold, Laemmli *et al.*, 1978). It suggests that DNA may have a structural role in stabilizing the basic shape of the chromosome, and that this is not solely the function of protein (Hadlaczky *et al.*, 1981b).

Until recently, plant material was not involved in such studies, because of the lack of an appropriate quantity of the isolated plant chromosomes.

Recently, a mass isolation procedure has been developed for plant chromosomes (Hadlaczky *et al.*, 1983) which has opened the way to involve plant material to such biochemical and structural studies of eukaryotic chromosomes (Hadlaczky *et al.*, 1982). The primary purpose of this article is to summarize the introductory data obtained from the study of protein-depleted plant chromosomes. In addition, I intend to bring together and evaluate the facts from animal and plant systems which allow certain deductions about eukaryotic chromosome structure.

Regarding the material presented here, it should be noted that experiments on protein-depleted plant chromosomes have been performed so far only with two species (i.e., *Papaver somniferum* and *Triticum monococcum*), and results concerning their structure came solely from our laboratory. It involves the risk of the crudeness of conclusions and certain partiality in the interpretation of results.

Although premature conclusions can initiate more sophisticated experiments and promote further progress, I have made every endeavour to avoid such shortcomings.

II. Structure of Protein-Depleted Chromosomes of Plants

The term "protein-depleted chromosomes" defines residual structures of isolated chromosomes after a certain deproteinization process. Deproteinization is a partial removal of certain classes of chromosomal proteins, therefore, protein-depleted chromosomes contain detectable amounts of residual proteins. In consequence, examination of the structure of protein-depleted chromosomes by light and EM is always coupled with a quantitative gel electrophoretic analysis of solubilized (extracted) and residual proteins associated with DNA.

"Whole-mount" preparations of intact metaphase chromosomes spread from mitotic protoplasts and isolated metaphase chromosomes of plants show a typical "folded-fiber" (Du Praw, 1965) appearance examined by transmission and scanning electron microscopy. Regardless of the species studied the average size of the highly coiled chromatin fibers in surface spread intact chromosomes is ~ 280 Å in diameter. Despite this uniform morphology, isolated chromosomes of plants show species-specific sensitivity to protein depletion (i.e., salt-extraction) treatment. Poppy chromosomes exhibit much higher structural stability than wheat chromosomes. In other words, nonhistone proteins of wheat chromosomes are less tightly bound to the DNA than nonhistones of isolated chromosomes of poppy. Even in a salt-free Tris buffer the majority of nonhistone proteins of wheat chromosomes can be solubilized and divalent cations have little effect in protecting the structural integrity of chromosomes. This difference between wheat and poppy chromosomes cannot be regarded as a distinction of monocotyledonous and dicotyledonous plants since the dicotyledonous broad bean (*Vicia faba*) chromosomes show almost identical sensitivity as wheat chromosomes (Hadlaczky and Praznovszky, unpublished).

In spite of the above dissimilarity of plant chromosomes, some general regularities of the plant chromosome organization can be established from protein depletion experiments. Furthermore, the different response of the chromosomal proteins of wheat provides additional evidence in elucidating the role of nonhistones in chromosome architecture. For experimental details regarding the protein-depleted chromosomes see Hadlaczky *et al.* (1982).

A. The Role of Nonhistone Proteins in Chromosome Structure of Plants

A progressive salt-extraction of plant chromosomes (i.e., treatment of chromosomes with increasing salt concentration) causes a gradual solubilization of chromosomal proteins. Extraction of chromosomes by 0.6 M salt results in a specific removal of H1 histone subfractions and some nonhistone proteins, while certain nonhistone proteins and all of the core histones are retained in the chromosomes.

The 2 M salt-treated chromosomes are composed of nonhistone proteins and DNA. With direct protein gel electrophoresis or subsequent acid extraction of protein-depleted chromosomes no detectable amount of histones can be found (Fig. 1). However, with 2 M salt treatment of poppy chromosomes, individual structures can be produced retaining the typical morphology of metaphase chromosomes. In spread EM preparation they appear as dense chromosome-shaped bodies; occasionally the separated chromatids can be seen. The majority of the DNA spreads out from the protein-depleted chromosomes forming a halo (Fig. 2).

In conclusion, it is obvious that histones are not required in maintaining the chromosome morphology, and it is reasonable to assume that nonhistone proteins play a key role in stabilizing the chromosome structure.

Experiments on protein-depleted wheat chromosomes provide strong support to this assumption. After extraction of wheat chromosomes with 0.6 M sodium chloride all of the core histones and only a small amount of some nonhistones remain associated with the DNA (Fig. 3). These "H1 and nonhistone-depleted

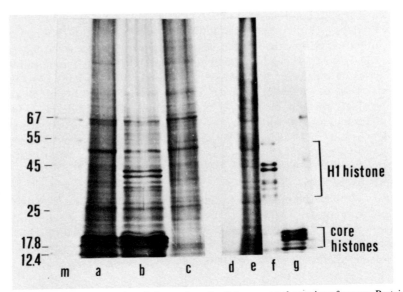

Fig. 1.   SDS polyacrylamide gel electrophoresis of chromosomal proteins of poppy. Proteins of
untreated chromosomes (a); extracted with 2 *M* sodium chloride (b); remaining in 2 *M* salt-treated
chromosomes (c); extracted from 2 *M* salt-treated chromosomes by subsequent perchloric acid
treatment (d); of chromosomes after 2 *M* sodium chloride and perchloric acid treatment (e). (f) H1
histone of poppy; (g) core histones of poppy; (m) molecular weight markers: bovine serum albumin
(67,000); pig brain tubulin (55,000); ovalbumine (45,000); chymotrypsinogen (25,000); myoglobin
(17,800); and cytochrom *c* (12,400). From Hadlaczky *et al.* (1982), with permission of Springer-
Verlag.

chromosomes'' are structurally unstable and even very weak mechanical forces
(e.g., low speed centrifugal force, surface tension at spreading etc.) cause an
immediate disintegration. Taking into consideration the effect of Tris buffer it is
very improbable that the absence of H1 histone is responsible for the lack of
structural stability. This is supported by the fact that wheat chromosomes incubated in salt-free Tris buffer retaining all of the H1 and core histones but only a
small amount of nonhistones (Fig. 3), show somehow similar structural lability.

From these observations it can be concluded that DNA and histones are insuf-
ficient to stabilize or preserve the chromosome structure in the absence of nonhistone proteins. Consequently, nonhistone proteins seem to play a fundamental
role in structural organization of plant chromosomes.

B. THE MORPHOLOGY OF PROTEIN-DEPLETED CHROMOSOMES OF PLANTS

Low salt extraction of poppy chromosomes with 0.6 *M* sodium chloride can
remove the majority of H1 histones and some nonhistone proteins from the

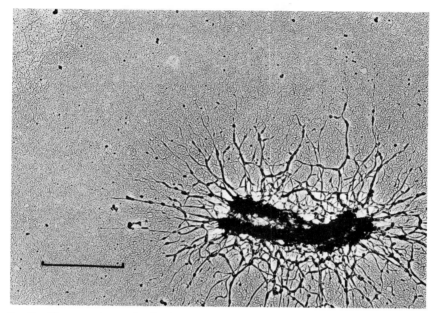

FIG. 2. Electron micrograph of a 2 *M* sodium chloride-extracted poppy chromosomes. Bar represents 2 μm.

chromosomes. Salt-treated chromosomes (0.6 *M*) containing all of the core histones and a considerable amount of nonhistones retain their typical morphology. Under EM poppy chromosomes extracted with low salt appear as compact chromosome-shaped bodies surrounded by a halo of DNA (nucleoprotein) loops (Fig. 4). Their EM appearance is identical with 2 *M* salt-extracted chromosomes though they contain only some nonhistone proteins (Fig. 5). The compact central body of the 2 *M* salt-extracted chromosomes can be loosened by salt-sucrose treatment and by reducing the concentration of protein-depleted chromosomes (DNA) at spreading.

The presence of sucrose in the 2 *M* salt extraction medium is presumed to prevent the aggregation of nonhistone proteins (Hadlaczky *et al.*, 1981b) without removing any extra proteins. On the other hand, a decrease in the concentration of protein-depleted chromosomes (DNA) increases the spreading force which is substantial in the transformation of structures into ''two-dimensional'' form. Sucrose-salt treatment of chromosomes and spreading at low DNA concentration produces looser, chromosome-shaped structures surrounded by a halo of DNA fibers. These central structures are composed of apparently aggregated DNA fibers and dispersed nucleoprotein particles. The centromeres appear as condensed double dots and similar structures can often be seen at the telomeres (Fig.

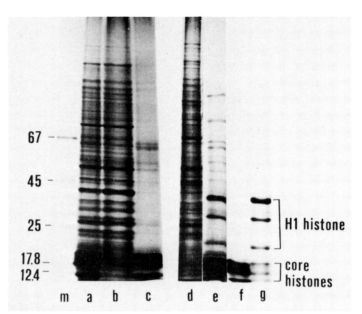

FIG. 3. SDS polyacrylamide gel electrophoresis of chromosomal proteins of wheat. Proteins of untreated chromosomes (a); extracted by 0.6 M sodium chloride treatment (b); remaining in chromosomes after 0.6 M salt treatment (c); extracted by 10 mM Tris treatment (d); remaining in chromosomes after Tris treatment (e); (f) core histones of wheat; (g) H1 histone of wheat; (m) molecular weight markers. From Hadlaczky *et al.* (1982), with permission of Springer-Verlag.

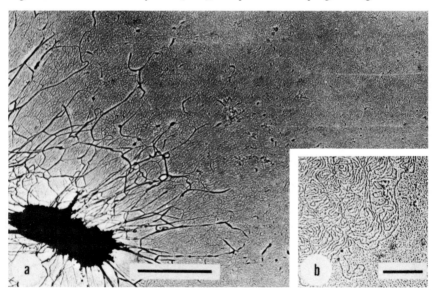

FIG. 4. Electron micrograph of an 0.6 M sodium chloride extracted chromosome of poppy (a). Bar represents 2 μm. (b) Enlarged part of the perimeter of DNA (nucleoprotein) halo. Bar represents 0.5 μm.

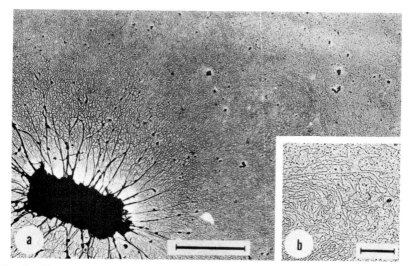

FIG. 5. The typical appearance of a 2 *M* sodium chloride extracted poppy chromosome under electron misroscope (a). Bar represents 2 μm. From Hadlaczky *et al.* (1982), with permission of Springer-Verlag. (b) Enlarged part of the perimeter of the DNA halo. Bar represents 0.5 μm.

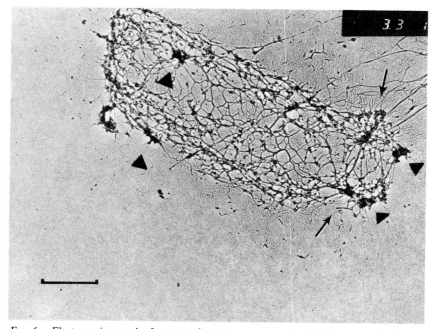

FIG. 6. Electron micrograph of a poppy chromosome extracted with 2 *M* sodium chloride in the presence of 5% sucrose and spread at a DNA concentration of 0.1 μg/cm². Arrows show centromeres and arrowheads indicate telomeres. Bar represents 2 μm. From Hadlaczky *et al.* (1982), with permission of Springer-Verlag.

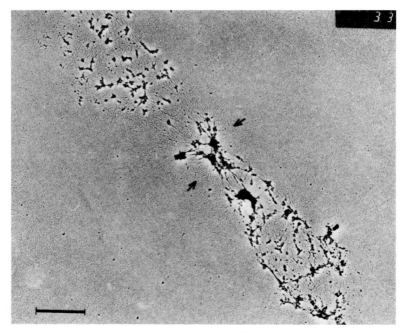

Fig. 7.   A part of a poppy chromosomes treated as in Fig. 6 but spread at a DNA concentration of
0.01 μg/cm². The shape of chromosome is indicated only by nucleoprotein particles and by residual
centromeres (arrows). Bar represents 2 μm. From Hadlaczky *et al.* (1982), with permission of
Springer-Verlag.

6). It strongly suggests that the structural (residual) nonhistones of protein-
depleted chromosomes are located in residual centromeres, telomeres, and in
discrete nucleoprotein particles. A further confirmation of this assumption can be
obtained from EM preparations made by extra dilution of protein-depleted chromosomes before spreading. In such preparations the thick ''scaffold-fibers'' are
also spread and the shape of the chromosome is indicated only by discrete
particles connected with DNA fibers and by condensed centromeric structures
(Fig. 7), or the shape of chromosomes is totally unrecognizable (Fig. 8) and only
dense paired structures (centromeres) are detectable in the DNA halo (Fig. 9).
This observation leads to the conclusion that centromeres are the most stable
structures of poppy chromosomes and in protein-depleted chromosomes they are
the ultimate detectable structural entities.

Experiments on salt-treated wheat chromosomes show that this conclusion can
be extended to wheat chromosomes as well. In stained light microscopic preparations of 0.6 *M* salt-treated wheat chromosomes only centromeres preserve their
condensed structure (Fig. 10). Moreover, *in situ* protein-depleted wheat chromosomes (salt treatment of chromosomes attached on EM grid) extracted with 1 *M*

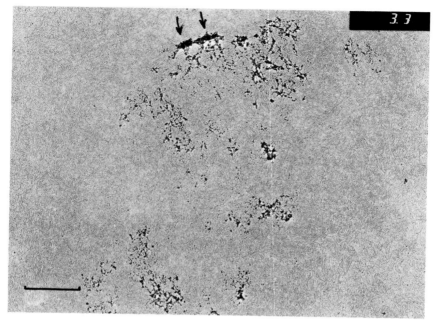

Fɪɢ. 8. Part of a protein-depleted poppy chromosomes treated as in Fig. 7. Arrows indicate the putative centromere. Bar represents 2 μm. From Hadlaczky *et al.* (1982), with permission of Springer-Verlag.

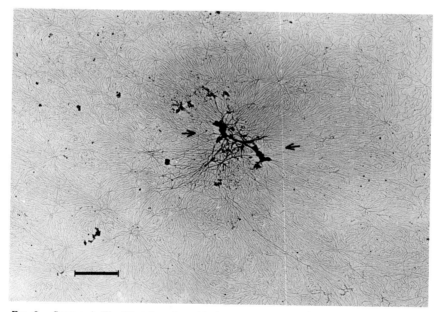

Fɪɢ. 9. Same as in Fig. 8 but the only residual structure is the putative centromere (arrows). Bar represents 1 μm. From Hadlaczky *et al.* (1982), with permission of Springer-Verlag.

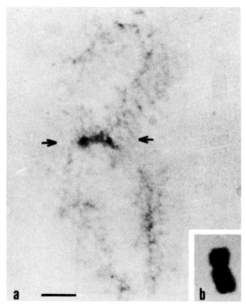

Fɪɢ 10.   Light micrographs of wheat chromosomes: (a) 0.6 M salt-extracted chromosome after Giemsa staining. Arrows show condensed centromere. Bar represents 5 μm. (b) Untreated chromosome at the same magnification. From Hadlaczky et al. (1982), with permission of Springer-Verlag.

sodium chloride produce loose chromosome-shaped structures where centromeres appear as condensed knobs on each of the chromatids (Fig. 11).

III. The Potential Use of Human Autoantibodies to Nuclear Antigens in Structural Studies of Plant Chromosomes

A wide variety of antibodies to nuclear antigens have been used in studying animal chromosomes by immunological methods. For example, antibodies specific for nucleosides (Freeman et al., 1971; Dev et al., 1972; Schreck et al., 1973), antisera to histones (Desai et al., 1972; Bustin et al., 1976), human autoantibodies of sera from patients suffering from autoimmune diseases or cancer (Krooth et al., 1961; Stenman et al., 1975; Moroi et al., 1980a), and monoclonal antibodies to chromatin components (Turner, 1981, 1982) have been successfully used in chromosome research. Human autoantibodies to nuclear antigens (reviewed by Tan, 1982) provide useful tools to study the structure and function of the chromosomal constituents. The question arises, whether chromosomes of evolutionary distant species contain uniform antigen determinants. If

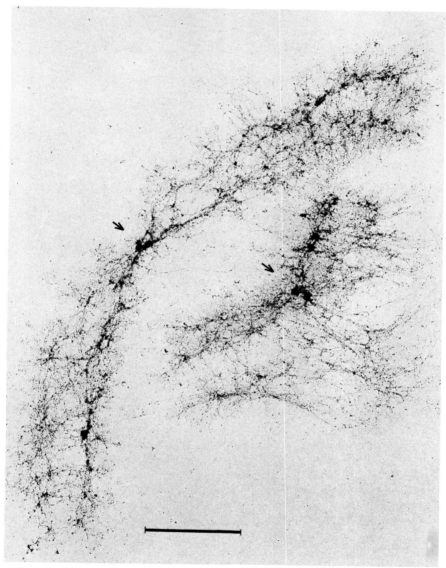

Fig. 11. Electron micrograph of wheat chromosomes extracted *in situ* with 1.0 *M* sodium chloride. Arrows show centromeres; bar represents 5 μm. From Hadlaczky *et al.* (1982), with permission of Springer-Verlag.

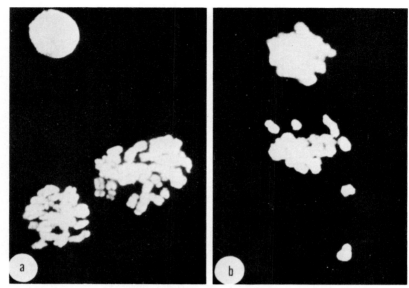

FIG. 12.   Indirect immunofluorescence of hamster (a) and poppy (b) chromosomes produced by serum from a patient with systemic lupus erythematosus.

so, and the antigen determinants are associated with structural elements it would be a strong indication for certain structural uniformity of eukaryotic chromosomes. The use of human autoantibodies in plant chromosome studies offers a unique opportunity for analysis of the ubiquitous macromolecules and highly conserved components of eukaryotic chromosomes.

The reality of this idea is supported by the fact that rabbit antibodies raised against chicken erythrocyte nucleosomes showed similar reactivity to nucleosomes of mammalian, insect, and plant chromosomes (Einck *et al.*, 1982).

In our laboratory surveying for human antinuclear autoantibodies showing identical specificity to antigen determinants of plant and animal chromosomes a number of sera have been tested. Some of the sera showed exactly the same reaction detected by indirect immunofluorescence on mitotic cells, isolated chromosomes, and nuclei of plant and animal material. Serum of a patient with systemic lupus erythematosus (SLE) contains high titer of antibody which reacts in the same fashion to plant and animal chromosomes and nuclei (Fig. 12). Results of the preliminary characterization of serum suggest that this antibody specific to the histone–DNA complex, therefore, it is assumed to be anti-nucleosome antibody (Hadlaczky and Praznovszky, 1985). Another group of sera from scleroderma patients exhibited strong uniform nucleolar and NOR (nucleolus organizer region of chromosomes) specific activity on plant and animal material (Fig. 13). It remains to be elucidated what kinds of macromolecules

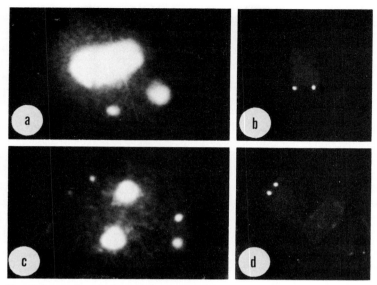

FIG. 13. Patterns of nucleolar staining produced by a serum from a patient suffering from scleroderma. (a) Isolated hamster nucleus showing bright nucleolus staining; (b) isolated nucleolar chromosome of hamster; (c) isolated poppy nucleus; (d) idolated poppy chromosomes, one of them showing bright NOR fluorescence.

serve as antigenic determinants in these cases, but in any event from these two examples it is obvious that some constituents of eukaryotic chromosomes remain immunologically conserved during evolution.

The use of these sera offers a new way to isolate and purify general components of the eukaryotic chromosomes, and makes possible their detailed structural and biochemical study. In addition, the present technology of molecular biology allows investigation of their possible functional role as well.

IV. Comparison of Structural Organization of Metaphase Chromosomes of Plants and Animals (Vertebrates)

As a summary of the presented data it is possible to make a final conclusion by suggesting a newer version of one chromosome model among many. Instead, I suppose a brief summary and discussion of those structural features of plant and animal chromosomes which seem to be general in higher eukaryotes is more profitable to the reader. At the first structural level of hierarchy in chromatin organization the basic repetitive unit is the nucleosome in both plants and animals. Although, certain biochemical alterations exist (for example, different histone composition of nucleosomes) the nucleosomes of plants and animals are formed on the same principles and organized into the familiar "beads on a string" structure of nucleosome fiber (for a discussion see Nagl in this volume). The next level of packing producing the 250–300 Å thick chromatin fiber seems to be also common in plant and animal chromosome at least at the level of EM morphology (Hadlaczky *et al.*, 1982). The question of the arrangement of nucleosomes, whether they form a coiled superhelix called solenoid (Finch and Klug, 1976; McGhee *et al.*, 1980) or are packed into superbeads (Hozier *et al.*, 1977; Rattner and Hamkalo, 1978), is unsettled and to make universal anyone of these models needs direct evidence. Assumptions for the existence of a higher level in the coiling of the chromatin fiber forming a "supersolenoid" (Bak *et al.*, 1977), a 2000 Å tube (Sedat and Manuelidis, 1978), or 500 Å projection (Marsden and Laemmli, 1979) seem to be more controversial and require more experimental support to accept.

Protein-depleted metaphase chromosomes of all species studied so far have some general characteristics regarding their structural organizations. Comparing the results of experiments made on protein-depleted plant and animal chromosomes some general conclusions can be drawn.

Removal of virtually all the histones and many nonhistone proteins from isolated chromosomes of human (HeLa) (Maio and Schildkraut, 1967; Adolph *et al.*, 1977), hamster (Maio and Schildkraut, 1967; Stubblefield and Wray, 1974; Jeppesen *et al.*, 1978; Okada and Comings, 1980; Hadlaczky *et al.*, 1981a), or plant (Papaver) cells (Hadlaczky *et al.*, 1982) leaves chromosome-shaped structures.

Despite the presence of all core histones in salt-treated chromosomes of wheat in the absence of nonhistone proteins the structural integrity of chromosomes is not preserved. These independent investigations result in the uniform conclusion that the structural role of histones is confined to the nucleosome fiber, whereas nonhistone proteins are the major structural elements in the higher ordered organization of chromosomes. The manifestation of the nonhistone proteins as an independent structural entity in the metaphase chromosomes is controversial. The "scaffold" or "scaffolding model" (Laemmli *et al.*, 1978) deduced from studies on protein-depleted human chromosomes postulates an axial proteinaceous core (i.e., nonhistone protein scaffold) at full length of each chromatid. The existence of such a distinct, stable entity in intact or protein-depleted chromosomes, on the basis of diverse experimental approaches, has been seriously queried even in the case of mammalian chromosomes (Goyanes *et al.*, 1980; Okada and Comings, 1980; Hadlaczky *et al.*, 1981b). Therefore, extending the model to plant chromosomes or to a wider field seems to be unfounded.

Notwithstanding, protein-depleted plant and animal chromosomes show some

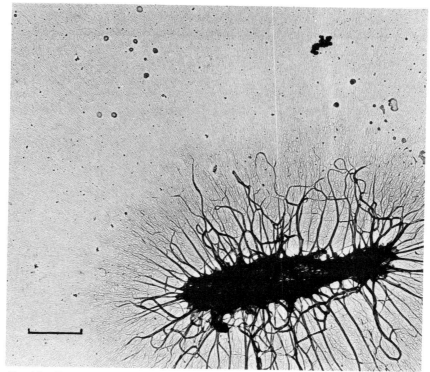

FIG. 14. Electron micrograph of a 2 *M* salt-extracted hamster chromosome. Bar represents 1 μm.

remarkable structural similarities; 2 *M* salt-extracted poppy and hamster chromosomes exhibit identical EM morphology showing dense chromosome-like structures surrounded by a DNA halo (Fig. 14, for a comparison see Figs. 2 and 5). Similarly, sucrose-salt treated plant and animal chromosomes prepared in the same way are hardly distinguishable (Fig. 15, compare with Fig. 6). Also, observations in somehow different experiments on plant (sucrose-salt treated spread at low DNA concentration) and animal chromosomes (sucrose-salt treated, spread at low DNA concentration, *in situ* extracted, DNase digested; Hadlaczky *et al.*, 1981b) led to the same conclusion that structural nonhistone proteins form nucleoprotein particles with certain DNA fragments. This conclusion incorporates the presumption of the structural role of certain DNA fragments in both animal and plant chromosome structure. Despite the accumulating information about the residual DNA of protein-depleted animal chromosomes (Razin *et al.*, 1978, 1979; Jeppesen and Bankier, 1979; Bowen, 1981; Kuo, 1982) this

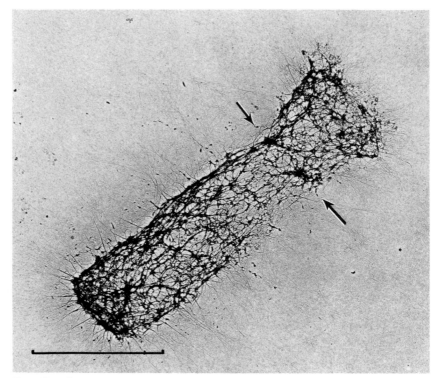

FIG. 15.   Electron micrograph of a hamster chromosome treated as poppy chromosome in Fig. 6.
Bar represents 5 μm.

presumed structural role of DNA remains to be confirmed, especially as no
experimental data have been published regarding the residual DNA of plant
chromosomes.

Considering that a substantial feature of structural proteins of chromosomes is
their resistance to extraction procedures it is not merely speculation to assume
that "resistant" chromosomal structures may have structural importance too.
The most resistant structural entities of animal and plant chromosomes are the
centromeres. (The term centromere is used for the region of the primary constriction, for a discussion see Rieder, 1982). In mouse chromosomes digested with
DNase I only centromere regions remain intact (Rattner *et al.*, 1978). In structurally well preserved protein-depleted hamster chromosomes, centromeres are
always visible as rounded compact structures; moreover, the sole organized
structure in protein-depleted chromosomes extracted *in situ* is centromere (Fig.
16) (Hadlaczky *et al.*, 1981a,b). This is not a unique property of protein-depleted
hamster chromosomes since with an improved visualization of histone-depleted

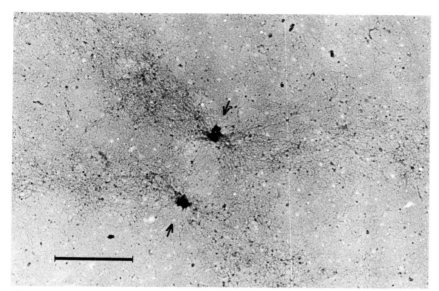

FIG. 16. Electron micrograph of a hamster chromosome extracted *in situ* with 2 *M* sodium chloride in the presence of 5% sucrose. Arrows show residual centromere; bar represents 2 μm. From Hadlaczky *et al.* (1981b), with permission of Springer-Verlag.

chromosomes Ernshaw and Laemmli (1983) have also managed to demonstrate the existence of residual centromeres in protein-depleted human (HeLa) chromosomes. Centromeres of protein-depleted plant chromosomes also possess identical structural stability under diverse experimental conditions (for a comparison see Figs. 6–11 and 15–16).

Considering the structural stability of centromeres in both plant and animal chromosomes we have suggested that centromeres might have structural importance in chromosome architecture (Hadlaczky *et al.*, 1982). It is tempting to assume that this presumptive structural role of centromeres is not limited to the centromeric region of chromosomes, and they have a more extensive structural significance. For example, in respect to chromosome condensation centromeres may serve as "foot-stones" providing a structural basis in the maintainance of the integrity of the individual chromosomes. In the light of this it is worth mentioning that centromere like structures can be seen in both plant (Church and Moens, 1976; Moens and Moens, 1981) and animal nuclei (Moroi *et al.*, 1980b; Brenner *et al.*, 1981) at all stages of cell cycle.

Preliminary immunological studies on intact plant and animal chromosomes show that they do contain structural entities, such as nucleosomes, and nucleolus organizing regions, which are identical respecting their immunological characteristics. The high specificity of immunological reactions strongly indicates that essential similarities might exist in eukaryotic chromosomes and diversity of evolutionary distant species does not inevitably pertain to these structures of prime importance.

Finally, on the basis of data accumulated so far, there is no experimental or theoretical objections to the conclusion that the basic principles of supramolecular organization of metaphase plant chromosomes established so far are the same as in animal chromosomes. These structural principles of metaphase chromosomes can be summarized as follows.

1. The primary structural level of chromosome organization is the formation of the nucleosomes and 100 Å nucleosome fiber showing certain morphological and immunological uniformity.

2. Hitherto, the unanimously accepted highest rate structural element in the metaphase chromosomes is the 250–300 Å chromatin fiber.

3. The direct structural role of histones is confined to the nucleosome fiber, hereby, to the chromatin fiber. Apart from this, histones have no verified contribution to the higher ordered organization of chromosomes.

4. Certain nonhistone proteins have crucial importance in the higher ordered structure of metaphase chromosomes. The highest structural complexity of these nonhistones detected in protein-depleted chromosomes appears in the form of residual centromeres. Their structural significance and that of DNA fragments associated with residual nonhistones is not elucidated.

It seems that structural nonhistone proteins are those chromosomal components which give the material basis of the higher ordered packing of the 300 Å chromatin fiber. Nevertheless, probably the most intriguing question, the mechanism of this packing together and its functional consequences, still remains to be answered. Since all of the existing models for the higher ordered organization of chromosomes must deal with serious objections, the early statement of Hearst and Botchan (1970) is the latest that we know: "additional molecules (i.e. nonhistone proteins) are added to the chromatin fiber at discrete spots and serve as glue for chromosomal folding."

ACKNOWLEDGMENTS

I thank Dr István Raskó and Mr. Brent Eddington for their critical reading of the manuscript and Miss Zsuzsanna Rácz for her patience in typing it.

REFERENCES

Adolph, K. W., Cheng, S. M., and Laemmli, U. K. (1977). *Cell* **12**, 805–816.
Bak, A. L., Zeuthen, J., and Crick, F. C. H. (1977). *Proc. Natl. Acad. Sci. U.S.A.* **74**, 1595–1599.

Bowen, B. C. (1981). *Nucleic Acids Res.* **9**, 5093–5108.
Brenner, S., Pepper, D., Berns, M. W., Tan, E., and Brinkley, B. R. (1981). *J. Cell Biol.* **91**, 95–102.
Bustin, M., Yamasaki, H., Goldblatt, D., Shani, M., Huberman, E., and Sachs, L. (1976). *Exp. Cell Res.* **97**, 440–444.
Church, K., and Moens, P. B. (1976). *Chromosoma* **56**, 249–263.
Desai, L. S., Pothier, L., Foley, G. E., and Adams, R. A. (1972). *Exp. Cell Res.* **70**, 468–471.
Dev, V. G., Warburton, D., Miller, O. J., Miller, D. A., Erlanger, B. F., and Beiser, S. M. (1972). *Exp. Cell Res.* **74**, 288–293.
Du Praw, E. J. (1965). *Proc. Natl. Acad. Sci. U.S.A.* **53**, 161–168.
Earnshaw, W. C., and Laemmli, U. K. (1983). *J. Cell Biol.* **96**, 84–93.
Einck, L., Dibble, R., Frado, L. L. Y., and Woodcock, C. L. F. (1982). *Exp. Cell Res.* **139**, 101–110.
Finch, J. T., and Klug, A. (1976). *Proc. Natl. Acad. Sci. U.S.A.* **73**, 1897–1901.
Freeman, M. V. R., Beiser, S. M., Erlanger, B. F., and Miller, O. J. (1971). *Exp. Cell Res.* **69**, 345–355.
Goyanes, V. J., Matsui, S., and Sandberg, A. A. (1980). *Chromosoma* **78**, 123–135.
Hadlaczky, Gy., and Praznovszky, T. (1985). In preparation.
Hadlaczky, Gy., Sumner, A. T., and Ross, A. (1981a). *Chromosoma* **81**, 537–555.
Hadlaczky, Gy., Sumner, A. T., and Ross, A. (1981b). *Chromosoma* **81**, 557–567.
Hadlaczky, Gy., Praznovszky, T., and Bisztray, Gy. (1982). *Chromosoma* **86**, 643–659.
Hadlaczky, Gy., Bisztray, Gy., Praznovszky, T., and Dudits, D. (1983). *Planta* **157**, 278–285.
Hearst, J. E., and Botchan, M. (1970). *Annu. Rev. Biochem.* **39**, 151–182.
Hozier, J., Renz, M., and Nehls, P. (1977). *Chromosoma* **62**, 301–317.
Jeppesen, P. G. N., and Bankier, A. T. (1979). *Nucleic Acids Res.* **7**, 49–67.
Jeppesen, P. G. N., Bankier, A. T., and Sanders, L. (1978). *Exp. Cell Res.* **115**, 293–305.
Krooth, R. S., Tobie, J. E., Tjio, J. H., and Goodman, H. C. (1961). *Science* **134**, 284–286.
Kuo, M. T. (1982). *J. Cell Biol.* **93**, 278–284.
Laemmli, U. K., Cheng, S. M., Adolph, K. W., Paulson, J. R., Brown, J. A., and Baumbach, W. R. (1978). *Cold Spring Harbor Symp. Quant. Biol.* **42**, 109–118.
Mace, M. L., Jr., Daskal, Y., Busch, H., Wray, V. P., and Wray, W. (1977). *Cytobios* **19**, 27–40.
McGhee, J. D., Rau, D. C., Charney, E., and Felsenfeld, G. (1980). *Cell* **22**, 87–96.
Maio, J. J., and Schildkraut, C. L. (1966). *Methods Cell Physiol.* **2**, 113–130.
Maio, J. J., and Schildkraut, C. L. (1967). *J. Mol. Biol.* **24**, 29–39.
Marsden, M. P. F., and Laemmli, U. K. (1979). *Cell* **17**, 849–858.
Moens, P. B., and Moens, T. (1981). *J. Ultrastruct. Res.* **75**, 131–141.
Moroi, Y., Peebles, C., Fritzler, M. J., Steigerwald, J., and Tan, E. M. (1980a). *Proc. Natl. Acad. Sci. U.S.A.* **77**, 1627–1631.
Moroi, Y., Hartmann, A. K., Nakame, P. L., and Tan, E. M. (1980b). *J. Cell Biol.* **90**, 254–259.
Mullinger, A. M., and Johnson, R. T. (1979). *J. Cell Sci.* **38**, 369–389.
Mullinger, A. M., and Johnson, R. T. (1980). *J. Cell Sci.* **46**, 61–86.
Okada, T. A., and Comings, D. E. (1980). *Am. J. Hum. Genet.* **32**, 814–832.
Paulson, J. R., and Laemmli, U. K. (1977). *Cell* **12**, 817–828.
Rattner, J. B., and Hamkalo, B. A. (1978). *Chromosoma* **69**, 373–379.
Rattner, J. B., Branch, A., and Hamkalo, B. A. (1975). *Chromosoma* **52**, 329–338.
Rattner, J. B., Krystal, G., and Hamkalo, B. A. (1978). *Chromosoma* **66**, 259–268.
Razin, S. V., Mantieva, V. L., and Georgiev, G. P. (1978). *Nucleic Acids Res.* **5**, 4737–4751.
Razin, S. V., Mantieva, V. L., and Georgiev, G. P. (1979). *Nucleic Acids Res.* **7**, 1713–1735.
Rieder, C. L. (1982). *Int. Rev. Cytol.* **79**, 1–58.
Schreck, R. R., Warburton, D., Miller, O. J., Beiser, S. M., and Erlanger, B. F. (1973). *Proc. Natl. Acad. Sci. U.S.A.* **70**, 804–807.

Sedat, J., and Manuelidis, L. (1978). *Cold Spring Harbor Symp. Quant. Biol.* **42,** 331–350.
Stenman, S., Rosenqvist, M., and Ringertz, N. R. (1975). *Exp. Cell Res.* **90,** 87–94.
Stubblefield, E., and Wray, W. (1974). *Cold Spring Harbor Symp. Quant. Biol.* **38,** 835–843.
Tan, E. M. (1982). *Adv. Immunol.* **33,** 167–240.
Turner, B. M. (1981). *Eur. J. Cell Biol.* **24,** 266–274.
Turner, B. M. (1982). *Chromosoma* **87,** 345–357.

INTERNATIONAL REVIEW OF CYTOLOGY, VOL. 94

The Kinetochore

M. B. E. Godward

*School of Biological Sciences, Queen Mary College, University of London,
London, England*

I. Introduction

A. Use of Terms

It is to be expected that the extensive literature will be full of numerous terms applied to the same object. The terms used here are thought to be self-explanatory and are some of those in common use. However, it may be as well to say that the terms "polar organelle" and "spindle pole body" are used for structures, usually electron-dense or associated with electron-dense material, which are situated at the poles of the mitotic spindle at some stage in their developmental cycle. The terms "kinetochore" and "centromere" will be discussed; but "kinetochore" is used for the structure seen with the electron microscope, "centromere" in a theoretical sense.

B. Definition of the Centromere

The term "centromere" was adopted by Darlington (Darlington, 1936), the term "kinetochore" by Sharp (Sharp, 1934), for the characteristic body forming part of the chromosome at its spindle attachment (Darlington, 1937). Other terms in use at that time and earlier are cited by Darlington (1937, p. 536). Later references will be found in Reider (Reider, 1982).

77

C. Limitation of the Topic

No attempt is made to consider mechanisms of mitosis. Only the nature of the kinetochore (centromere) and of the diffuse kinetochore is considered, although the mitotic apparatus insofar as it might throw light on the above topics is reviewed.

II. Evolution of the Ultrastructurally Visible Kinetochore

A. Survey of Evidence

1. The "Trilaminar" Kinetochore

The trilaminar kinetochore is characteristically that of mammals (Roos, 1973; Rieder, 1982); in section it has a dark/light/dark stratification; it appears as a projecting layer on the chromatin, with attached microtubules. Three hypothetical possibilities for its fibrillar organization have recently been put forward (Roos, 1977). However, there is some variation in appearance when the whole range of organisms recorded with trilaminar kinetochores is surveyed; there is also considerable variation in their position in the nucleus. They are in general found in the animal kingdom as well as in some of the lower eukaryotes both plant and animal; a selection includes *Euglena* (Gillot and Triener, 1978), *Hantzschia* (Pickett-Heaps *et al.*, 1982), *Hydrodictyon* (Marchant and Pickett-Heaps, 1970) *Microspora* Pickett-Heaps, *Chlamydomonas* (Plaskitt and Davies, 1970), *Oedogonium* (Pickett-Heaps, and Fowke, 1969, 1970), *Cladophora* (Mughal and Godward, 1973), *Spirogyra* (Mughal and Godward, 1973), *Membranoptera* (McDonald, 1972), *Apoglossum* (Dave, 1979), *Polysiphonia* (Scott *et al.*, 1980), *Eimeria* (Dubremetz, 1973), *Minchinia* (Perkins, 1975), *Barbulanympha* (Hollande and Valentin, 1968), *Trichonympha* (Kubai, 1973), and *Nyctotherus* (Eichenlaub-Ritter and Ruthman, 1982). Some points of interest concerning kinetochores in mainly lower eukaryotes, which are thought to have a bearing on the status of kinetochores generally, are considered, together with other relevant features of mitosis in these organisms.

Polar organelles are included in this survey, since they are in different organisms associated with the nuclear envelope and with microtubules, both structures associated with kinetochores. Polar organelles may, in different organisms, display a stratified structure not unlike that of the trilaminar kinetochore.

2. Kinetochore in the Nuclear Envelope

This occurs in the Hypermastigotes *Barbulanympha* (Hollande and Valentin, 1968) (Fig. 1), *Trichonympha* (Fig. 2) (Kubai, 1973), and the possible Dinoflagellate *Syndinium* (Fig. 3) (Ris and Kubai, 1974). In the dinoflagellate *Amphidinium* (Oakley and Dodge, 1976a) the position of the microtubule attachment point is the same, although only a small electron-dense pad represents the kinetochore. In *Trichonympha* (Fig. 2), however, in the kinetochore itself there is an outer rather electron-dense layer, then a very dark layer in line with the nuclear envelope, then an inner less dense layer, to which the fibrils of chromatin are attached. Microtubules are attached on the outside layer. In *Barbulanympha* (Fig. 1) there is a rather dense outer layer, in line with the nuclear envelope, but considerably thicker; within this a complex inner layer projects into the nucleus. Microtubules are again attached on the outer layer only, the chromosome to the inner layer. In *Syndinium* (Fig. 3) the stratified kinetochores appear part of the nuclear membrane; they are concentrated below a pit in the nuclear envelope, which surrounds the centrioles. Microtubules connect the centrioles to the kinetochores, the chromosome being attached on the inner face of the kinetochore. No centrioles are involved in the other genera, although the organisms are flagellated, and have many flagellar bases.

These examples have been mentioned first because they establish the fact that kinetochores can be part of the nuclear envelope, with the chromosome of course attached to them in that position. The degree of laminated structure in these kinetochores can be extreme, or negligible; but in all cases, electron-dense material is present in the kinetochore.

3. *Polar Organelles and the Nuclear Envelope*

In *Syndinium* centrioles (basal bodies) are present as functioning spindle pole bodies (spb) or microtubule organizing centers (mtoc) and it may be recalled that centrioles generally are frequently surrounded by electron-dense material (pericentriolar) in an amorphous cloud.

In some other organisms the spb is not a centriole but instead an electron-dense structure of variable form. In *Boletus* (McLaughlin, 1971) it has a round form vaguely reminiscent of one or two centrioles; it is closely applied to the nuclear envelope which beneath it is filled with electron-dense material. No kinetochore has been observed here. In *Uromyces* (Fig. 7) (Heath and Heath, 1976) it is a flat three or four layered structure, bipartite or simple and round, applied to (and part of?) the nuclear membrane. Microtubules are attached to it both inside the nuclear envelope and outside. Kinetochores are very small electron-dense areas on the chromatin and well within the nucleus. Other Basidiomycetous fungi are much like the above, although in Heterobasidiomycetous yeasts (McCully and Robinow, 1972) a spindle forms outside the nuclear envelope between the replicates of the spb and the envelope breaks down to let it into the nucleus, afterward repairing the break. In ascomycetes (Zickler, 1970; Lu, 1967) the spb may be an electron-dense layer applied in part to the outer membrane of the nuclear envelope and in part standing off from it, while the nuclear membrane itself is highly electron dense in the polar regions; microtubules are attached both inside to the nuclear envelope, and outside to the spb. In *Saccharomyces* (Fig. 5) (Robinow

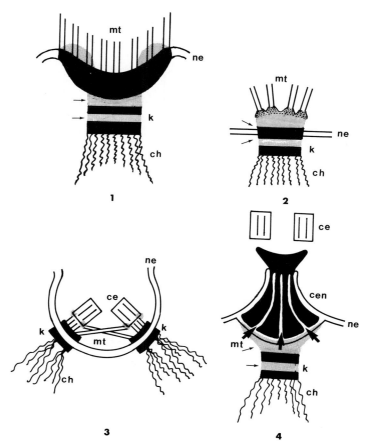

Figs. 1–9.   Kinetochores of various species: (1) *Barbulanympha ufalula* Cleveland (from Hollande and Valentin, 1968); (2) *Trichonympha agilis* (from Kubai, 1973); (3) *Syndinium* (from Ris and
Kubai, 1974); (4) *Eimeria necatrix* Johnson (from Dubremetz, 1973); (5) *Saccharomyces cerevisiae*
(from Byers and Goetsch, 1975); (6) *Fusarium oxysporum* (from Aist and Williams, 1972); (7)
Uromyces phaseoli var *vignae* (from Heath and Heath, 1976); (8) *Pinnularia maior* (from Pickett-
Heaps *et al.*, 1978); and (9) *Physarum polycephalum* (from Sakai and Shigenaga, 1972). ne, Nuclear
envelope; mt, microtubules; ch, chromosome; k, kinetochore; sp, single plaque; op, outer plaque; s,
satellite; spb, spindle hole body; nao, nucleus associated organelle; pp. polar plates; c, collar; ps,
primordium of spindle microtubules; ce, centriole; cen, centrocone. Black, very electron-dense
material; dotted, less dense material; electron density according to closeness of dots.

and Marak, 1966), there is no spb outside the nuclear envelope; only the nuclear
envelope itself is highly electron-dense at the poles, microtubules attaching to it
both inside and outside. Prior to nuclear division, a small "satellite" of electron-
dense material is developed on the cytoplasmic side, on the replicate and parent
polar electron-dense organelle (Byers and Groetsch, 1975); the movement of the

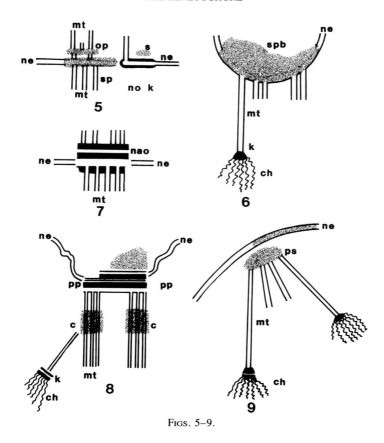

FIGS. 5–9.

replicate to the opposite pole may be compared with that of the kinetochores in the nuclear envelope of *Trichonympha,* where "sister kinetochores are separated and re-distributed on the nuclear surface". . . . "the observation of a gathering of a number of . . . kinetochores at each pole . . . is . . . suggestive" (Kubai, 1973).

In these examples, mainly fungi, we see the electron-dense material characteristically forming, or surrounding, the polar organelle, present also in the nuclear envelope to which it is closely applied, and which appears modified by its presence. An accompanying feature is the presence only of a very small kinetochore which when visible is located on the chromosome or noncondensed chromatin.

4. *Kinetochores Related to the Nuclear Envelope Although Not Part of It*

In the sporozoan *Eimeria* (Fig. 4) (Dubremetz, 1973) the spb is a "centrocone" of electron-dense material, interrupting the nuclear envelope; its base is stratified in appearance, and on it are formed the trilaminar kinetochores to which the chromosomes are attached. During mitosis, microtubules separate the kinetochores from the centrocone; but at the conclusion of division these microtubules shorten and disappear, and the kinetochores are said to ''melt'' back into the base of the centrocone.

Another interesting example is the fungus *Fusarium oxysporum* (Fig. 6) (Aist and Williams, 1972, Fig. 14) where a rather indistinctly stratified spb located on the nuclear envelope, at metaphase is in contact with the interior of the nucleus through polar fenestrae, through which a microtubule passes to a well-defined if small electron-dense kinetochore on the chromosome. The microtubule is not a very long one, and the resulting relationships of polar organelle, microtubule, and kinetochore are not far from those in *Eimeria,* at least at metaphase (Fig. 4).

5. *Intranuclear Polar Organelles*

So far the polar organelles considered have been situated on, or in, the nuclear envelope. There are examples in which they originate in, or persist at all times, inside the nucleus; this is the case in the plasmodium of *Physarum* (Fig. 9) (Sakai and Shigenaga, 1972; Tanaka, 1973) where an amorphous electron-dense body appears in the nucleus at prophase, apparently divides, and is then seen to diminish in size as the microtubular spindle is formed although some remains, pushed into near contact with the nuclear envelopes at the spindle poles. Other examples are the haplosporidan *Minchinia* species (Perkins, 1975) and others, where an electron-dense spindle pole body is always present at either end of a microtubular spindle, which lengthens during mitosis pushing the electron-dense material close to, but not in contact with, the nuclear envelope.

In the above instances, no clear electron-dense kinetochore structure appears in the electron micrographs (cf. Heath, 1980). They are examples, however, of the generation of electron-dense material within the nucleus, and the attachment of microtubules to it, although only polar attachment is obvious.

6. *Stratified or Laminar Polar Organelles*

Examples are seen in the fungus *Uromyces* (Fig. 7) (Heath and Heath 1976), slime mould *Polysphondylium* (Roos, 1975), and the polar complexes of the diatoms—*Lithodesmium* (Manton *et al.,* 1969), *Pinnularia* (Fig. 8) (Pickett-Heaps, 1978), and others (Pickett-Heaps *et al.,* 1982).

7. *Very Small Kinetochores*

Kinetochores small in relation to the bulk of the chromosome are seen in the diatom *Hantzschia* (Pickett-Heaps *et al.,* 1982) although most of these authors' diatom studies do not show this clearly trilaminar structure. Small kinetochores which could clearly carry only one or two microtubules are seen in *Physarum* (Fig. 9) (Sakai and Shigenaga, 1972), *Fusarium* (Aist and Williams, 1972), *Saprolegnia* (Heath and Greenwood, 1968), *Amphidinium* (Oakley and Dodge,

1974, 1976b), and *Uromyces* (Heath and Heath, 1976) to quote some examples. Others have already been referred to.

8. *No Kinetochores*

A list could be made; some examples are *Boletus, Saccaromyces, Catenaria* (Ichida and Fuller, 1968), *Blepharisma* (Jenkins, 1967) and *Amoeba* (Roth, 1967). However, one cannot say that kinetochores would never be found, in view of the difficulties of preparation and the ease with which they could be missed, unless the direct attachment of microtubule to chromatin fiber has been observed, as in *Saccharomyces* (Peterson and Ris, 1976).

B. Origin and Development of Kinetochores

The origin of kinetochores can be looked at from the points of view of ontogeny or phylogeny. To pick out individual structures and put them together to produce a kinetochore is perhaps the ontogenetic approach.

One possibility is that put forward by Roos (1975): "a chromosome microtubule terminates in amorphous or finely fibrillar material at a certain distance from the surface of the chromatid proper. One can consider these organelles linking a chromatid to a *single* microtubule as 'unit kinetochores' representing a primitive condition and peculiar to small chromosomes." He is discussing *Polysphondylium,* one of the small kinetochore types. One can go on to say that some organisms have not merely small chromosomes but noncondensed chromosomes with even less attachment space. It is, it appears to me, quite reasonable to suggest that a number of microtubules side by side, each with its basal bunch of amorphous or finely fibrillar material, would provide the outer lamina or perhaps two outer laminae of the trilaminar kinetochore.

On this view the trilaminar kinetochore would be an aggregate of microtubule bases, with their associated proteins forming the laminae (see Figs. 10–12).

A second possibility concerns the generation of electron-dense material, and in particular of stratified electron-dense material, in the form of multilaminar polar organelles or complexes. Although the molecular composition of these organelles is not precisely known, it is clearly of such a nature that combination in a lamellar form, with stratification of denser and less dense components, is not infrequent. It is also of such a nature as to attach, or nucleate, microtubules from subunits. Its distribution in the cell, taking into account the various organisms studied, covers the following range: (1) outside the nuclear envelope, (2) in the nuclear envelope, (3) in a polar fenestra of the nuclear envelope, (4) in the nucleus next to the nuclear envelope, (5) in the interior of the nucleus, to which one might plausibly add (6) in the nucleus on the chromosome as the electron microscopically visible kinetochore. The situations (2) and (4) above refer to recognized kinetochores as well as to polar organelles.

In some instances there is uncertainty about a laminar structure in the nu-

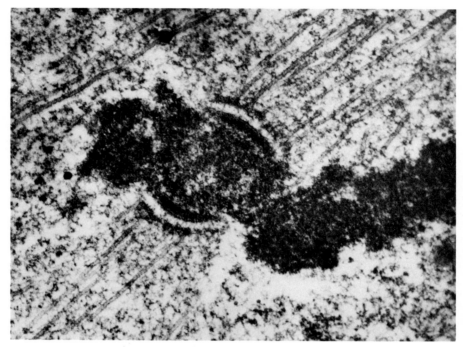

Fig. 10. *Oedogonium cardiacum* (from Pickett-Heaps and Fowke, 1969). Mitotic metaphase, ×40,000. Sister-kinetochores on opposite sides of the chromosome, showing trilaminar structure, the outermost dark layer formed by conical bases to the microtubules.

cleus—as in the species of *Trypanosoma* (Vickerman and Preston, 1970) where "laminated plaques" are present in the spindle but chromatin associations are not observed since the chromatin is not condensed sufficiently.

Whether, in all these situations the same molecules or molecules with similar properties are present in all the organisms cannot be said at present, but it is a possibility that protein molecules of a type liable to formation of laminated structures are available in the nucleus, are of assistance in the nucleation of microtubules, and are able to be associated with chromosomal DNA as well as with microtubules. If these generally available molecules in some organisms attach to the centromeric region and produce a trilaminar kinetochore, it follows that one need not regard this structure as part of the chromosome. That it is not necessary that sufficient protein molecules required to associate with DNA to permit attachment of microtubules are present in higher plants and those lower organisms where no kinetochore or only a minute one exists seems evident.

Turning to the phylogenetic approach, the question may be asked, is the above series (1) to (6), or something like it, a linear series of evolutionary steps? We do

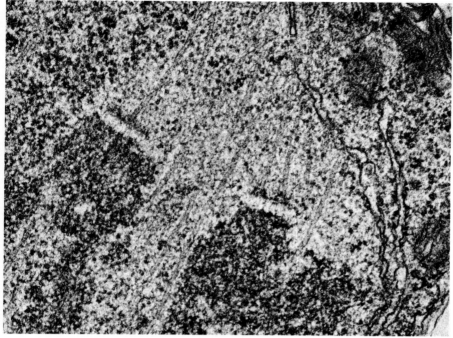

FIG. 11. *Polysiphonia harveyi* (from Scott *et al.*, 1980). Mitotic anaphase, trilaminar kinetochores showing structure essentially similar to Fig. 10 although on a smaller scale, ×80,000.

not see clear groupings of organisms which have taken one or more of these steps; instead we see the same step taken in totally unrelated as well as related organisms, for example, the chromosome-located trilaminar kinetochore. Yet where it is absent, as in most higher plants, it seems to have been lost, rather than not to have formed since it is present in the lower green plants (algae). On the other hand, a "ball-and-cup" type kinetochore, regarded as typical of higher plants, is said to be present in prophase of the rat kangaroo, differentiating into the trilaminar structure at prometaphase. In most fungi, where only a small simple kinetochore carrying one microtubule is present—or there may be apparent direct attachment of microtubule to chromatin—it seems unlikely that a trilaminar kinetochore has been lost.

These contrasted observations do not allow assessment of the trilaminar kinetochore as primitive or advanced.

It could be advocated that step (2), laminar kinetochore in the nuclear envelope, is restricted to Hypermastigotes and some Dinoflagellates or organisms possibly related to them, and that they are primitive in their kinetochore develop-

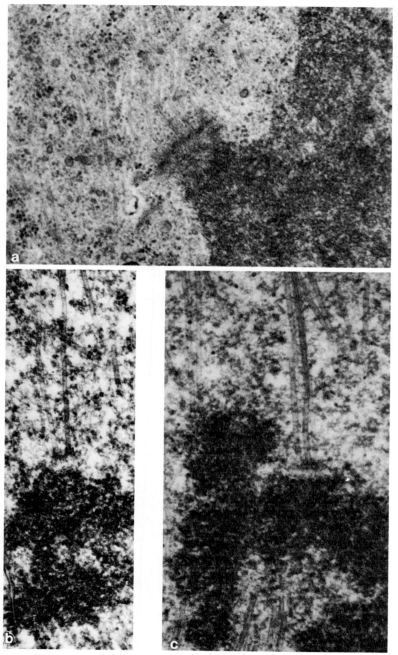

FIG. 12. (a) *Mnium hornum* from Lambert. ×57,000. Meiotic metaphase I, typical trilaminar kinetochore. (b,c) *Chlamydomonas reinhardii*, ×84,000. By permission of Roy Davis and K. Audrey Plaskitt, John Innes Institute. Typical trilaminar kinetochores with outer dark layer formed by bases of microtubules.

ment. Yet even here, a close approach, step (4), is seen in an unrelated Sporozoan—not necessarily disproof. It may be significant that in *Prococentrum micans* (Dinoflagellate) a mechanism of mitosis appears to be present that does not involve kinetochores or attached microtubules (Soyer, 1982). Possibly the little pad of electron-dense material on the outer membrane of the nuclear envelope of *Amphidinum* (Dinoflagellate) with its single attached microtubule, clearly connected to the chromatid within (Oakley and Dodge, 1974), represents an early stage in the utilization of microtubule-attaching proteins and the initiation of nuclear envelope-sited kinetochores.

The evolution of the kinetochore, both as to structure and location, no doubt depended on gene changes, some perhaps acting only indirectly. Given an original set of genes, certain changes would prove to be the most likely, viable, or merely harmless; these could have arisen several times and survived independently. Their expression in present-day organisms may, or may not, indicate phytogenetic relationships. Much more information, especially biochemical information about the proteins or other substances present in the electron-dense material of the polar organelles, the nuclear envelope, and the kinetochores is needed. Therefore, it may be premature to try to involve kinetochore characteristics with phylogenetic relationships, although some do suggest themselves.

Attention may be directed to the probable early evolutionary significance of the nuclear envelope; few would deny that the initiation of this structure established the eukaryotes.

The nuclear envelope still has a particular role in relation to the nucleation of microtubules, which may be illustrated by two examples. In *Amoeba* (Roth, 1967) the nuclear envelope fragments in mitosis; the fragments enter the spindle, lie parallel to the microtubules in anaphase, and remain with them until telophase when they change position and line up around the chromosomes. In the moss *Mnium* (Lambert, 1980, Fig. 3a) microtubules are attached to the outer membrane of the nuclear envelope, which is coated with electron-dense material embedding the microtubule bases. The nuclear envelope breaks up in late prophase; the fragments of nuclear envelope are entirely surrounded by microtubules. The author comments "electron-opaque material, whether on the nuclear envelope or the kinetochore, initiates nucleation of microtubules."

There may be some significance in the fact that in almost all the organisms hitherto considered and which have electron-dense material around spb, and in the nuclear envelope or in kinetochores situated there, the mitosis is totally intranuclear or open to the cytoplasm only at the poles, whereas in *Amoeba* and *Mnium* the nuclear envelope fragments at mitosis, but does not seem to have lost its microtubule-nucleating properties.

Evidence has recently been put forward (Dave and Godward, 1982) that the polar organelle of the red alga *Apoglossum* is developed from a nuclear pore, which becomes elevated on a conical projection above the rest of the nuclear envelope. Replication of this polar organelle and the movement of the replicate round the nuclear envelope to the opposite pole have been described for another red alga, *Polysiphonia* (Scott *et al.*, 1980). The ability of the nuclear envelope and nuclear pores to replicate, producing annulate lamellae, is well known. The polar organelle of *Apoglossum* seems to have rather limited significance, microtubules being mainly attached to the nuclear envelope and the trilaminar kinetochores, while electron-dense material is mainly deposited in the polar regions of the nuclear envelope—a feature which signals the start of prophase.

It is perhaps not too speculative to offer the suggestion that the nuclear envelope is the site where the microtubule-nucleating protein is produced, assembled, or at least where it is concentrated; that it may remain there, or in part transfer to an spb, or to the chromosome, where it may or may not assemble in quantity, or form a laminated structure, prior to the attachment of microtubules, or incorporating proteins already associated with the bases of microtubules.

Thus the early evolution of kinetochores in the nuclear envelope would appear reasonable as would the evolution there of polar organelles; but maybe not all kinetochores and not all polar organelles had the same origin or followed the same developmental path.

The close association of interphase chromatin with the nuclear envelope has long been known, and more recently binding of chromosomal loops to "the peripheral lamina" of the nucleus (Hancock and Hughes, 1982) has been publicized—but this association does not have any special relationship to kinetochore (centromere) regions.

C. MOLECULAR NATURE OF THE KINETOCHORE

We can now quote the work of Carbon (Bloom and Carbon, 1982) and collaborators on the centromeric DNA of *Saccharomyces cerevisiae*. Although it has for a long time been possible to map the yeast centromere genetically, little could be seen cytologically or with the electron microscope, since the chromosomes do not appear condensed in nuclear division (using present fixation methods).

With the use of restriction enzymes, and overlap hybridization, it has now been possible to isolate the DNA of the centromeric regions of two yeast chromosomes II and XI, and insert the isolates into plasmids. The plasmids became minichromosomes, were perpetuated in mitosis, and paired with segregation at meiosis. Sequencing of base pairs in the DNA of the centromeric region has been carried out, with the results (reproduced from Bloom and Carbon, 1982) shown below (Fig. 13). As a result, we have CEN 3 and CEN 11, the centromere genes on chromosomes III and XI, respectively, showing a high degree of homology, consisting of elements I and II, a sequence 93–94% of A-T only, element III, a sequence of about 250 bp, and element IV. On either side of these sequences, nucleosomal DNA with specific nuclease cutting sites between each nucleosome extend for a total length of 12–15 nucleosomal subunits. Beyond are the genes

CDC 10 (on chromosome III) and Met 14 on chromosome XII; other important genes concerned with genetic recombination are present. The authors point out, in their discussion, that only one microtubule can be attached to the chromatin fiber (per chromosome) in the yeast spindle (Peterson and Ris, 1976); therefore in yeast we have the simplest possible kinetochore, with no visible structural differentiation. It is clear that a DNA sequence of nothing but A-T, as is characteristic of the middle of CEN 3 and CEN 11, could not possibly code for proteins; therefore any proteins taking part in kinetochore substructure are not products of the CEN DNA but have become bound to it. Microtubule-associated proteins (MAP) can bind to DNA, but not tubulin. "The attachment of microtubules to chromatin must be mediated by other than tubulin or DNA. Specific centromere-binding proteins are to be looked for using the CEN sequences as probes. . . . This work should lead to in vitro reconstruction of the centromere–microtubule complex" (Bloom and Carbon, 1982).

The homology between the CEN sequences is in harmony with the observations of centromere aggregation, notorious in *Drosophila melanogaster,* and often seen in organisms which are not polytene, at least in interphase or even prophase (e.g., Lafontaine and Luck, 1980).

The presence of numerous cutting sites for nucleases, adjacent to the CEN sequences, throws light on the "splitting of the centromere" which results in the formation of the isochromosome and metacentrics from telocentrics.

Prior to 1982, there was ample demonstration of the presence of proteins in the trilaminar kinetochore. A recent paper (Rieder, 1979) showed that RNP was present in the inner lamina of the kinetochore of PtK$_1$ cells (rat kangaroo) and in his discussion the author reviews earlier work and evidence; he suggests that the kinetochore RNP is a universal component of microtubule-organizing centers. A curious report of work with serum from "Crest" scleroderma patients (Brenner, *et al.,* 1981) showed that the serum (immunoflourescent-labeled) stained kinetochores in the dividing nucleus and specific spots in the interphase nucleus, and concludes "the antibody-specific antigen(s) is a previously unrecognised component of the kinetochore regions." Further indication of the nature of kinetochore proteins is the finding of salt-resistant centromeric and telomeric structures in protein-depleted chromosomes (Hadlaczky *et al.,* 1981).

It has long been observable (Du Praw, 1966) that "spread" chromosomes seen under the electron microscope show no trilaminar or other special structure at the centromere region, but only rather fewer chromatin fibrils than other regions; in a recent illustration of this (Utsumi, 1982) the author has pointed out that preparatory methods commonly in use for air-drying extract histones which stabilize the superstructure of nucleosomes and may result in loosening the structure of the chromosome—whether this is a closely apposed assay of nucleosomes (Rattner *et al.,* 1978; Miller and Bakken, 1972) or a radial loop arrangement (Marsden and Laemmli, 1979). On the other hand when only low

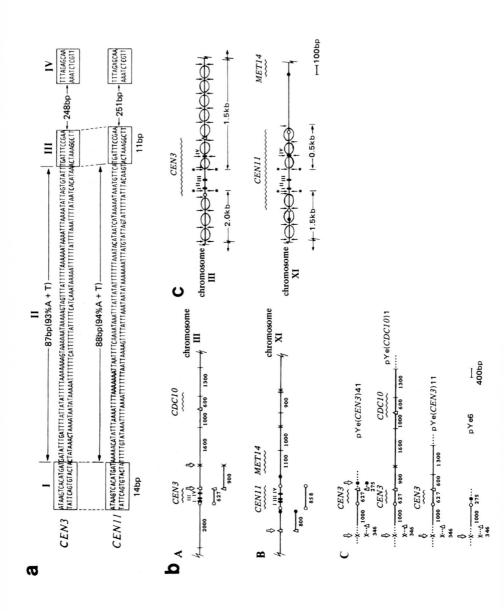

salt extraction is carried out, on a nuclease-digested chromosome, differentiated regions that derive from the kinetochore, rounded disc-like dark bodies, are found on the chromosome "scaffolds" (Earnshaw and Laemmli, 1983) of HeLa cells. Such "scaffolds" are regarded by many authors as probably random aggregations of displaced proteins (cf. Hadlaczky *et al.*, 1981; Burkholder, 1982; Rattner *et al.*, 1980)—possibly not relevant to the observations on the kinetochore by Earnshaw and Laemmli. One of the features of the kinetochore region of the air-dried spread chromosome is the fact that chromatin fibrils often connect the sister chromatids in this region, (Utsumi, 1982; Du Praw, 1966) although such fibrils also often connect the sister chromatids in other regions (Du Praw, 1966, Fig. 9.14).

FIG. 13.   Centromere genes on chromosomes III and XI of yeast (from Bloom and Carbon, 1982). (a) Spatial arrangement of sequence elements common to *CEN 3* and *CEN 11*. The regions of perfect sequence homology between *CEN 3* and *CEN 11*, elements I (14 bp), III (11 bp), and IV (10 bp), are positioned in an almost identical spatial arrangement within the two centromere sequences. The A + T-rich core segment (element II) lies between elements I and III in both DNA sequences. An overall homology of 71% is seen in the element I–II–III regions of *CEN3* and *CEN11* . The total length encompassed by sequence elements I–VI is 370 bp in *CEN3* and 374 bp in *CEN11* . (b) Comparative restriction maps of the centromere regions of yeast chromosomes III and XI and the relevant subcloned DNA fragments. The positions of the functional 627 bp *CEN3* DNA fragment on yeast chromosome III (A) and the 858 bp *CEN11* DNA fragment on yeast chromosome XI (B) are indicated. The regions of complete homology between the two centromeres, elements I, III, and IV, are indicated on *CEN3* (A) and *CEN11* (B). The very A + T-rich region (element II) lies between elements I and III in both centromeres. The relevant subcloned fragments are shown in (C). The solid lines indicate the yeast DNA inserts, and the dotted lines denote vector (pBR322) sequences. The complete structure of pYe(*CEN3*)41, pYe(*CDC*10)1, and pYe(*CEN3*)11 are given in Clarke and Carbon (1980). Plasmid r pYe6 is described in Fitzgerald-Hayes *et al.* (1982b). Only selected restriction sites are shown. The DNA fragments shown below each restriction map were used as hybridization probes; lengths are given in base pairs. The large arrows denote the restriction sites chosen for the nuclease mapping studies. Restriction enzyme cleavage sites are *Bam*HI (△), *Sau*3A (○), *Sal*I (●), *Hind*III (X), and *Eco*RI (|). (c) Schematic map of nuclease cleavage sites on centromeric chromatin. A map of the chromosomal region surrounding *CEN 3* and *Cen 11* shows the 627 bp *CEN 3* fragment, the 858 bp *CEN 11* fragment, and the relative position of the sequence elements I, III, and IV (see a) that are homologous and in identical spatial arrangements between *CEN 3* and *CEN 11*. The very A + T-rich region (element II) lies between elements I and III in both centromeres. The approximate position of the *MET* 14 gene on chromosome XI is shown (Fitzgerald-Hayes *et al.*, 1982a,b). Restriction enzyme sites are *Bam*HI (△). *Sau*3A (○), and *Sal*I (●). Microccal nuclease (↓) and DNase I (↑) cleavage sites on the centromeric chromatin are indicated. Asterisks indicate hypersensitive cleavage sites. The DNA fiber is presented in linear form to visualize the position of the 220–250 bp nuclease-resistant core and the nucleosomal subunits relative to the restriction maps of chromosomal DNA. In chromatin, the DNA is wrapped around the histone core to form the nucleosomal subunit (reviewed in McGhee and Felsenfeld, 1980). The conformation of DNA within the 220–250 bp nuclease-resistant core is not known. The arrows below the restriction maps indicate how far the ordered nucleosomal arrays extend from the nuclease-resistant centromere core.

That the kinetochore is able to bind tubulin, to nucleate microtubules, *in vivo* and *in vitro,* has been amply demonstrated (Snyder and McIntosh, 1975; De Brabander *et al.,* 1981); the field is reviewed and discussed in detail (Rieder, 1982). The same reviewer has also widely covered the chemistry of the kinetochore up to 1982.

D. REPLICATION

Now that the CEN DNA has been identified, centromere replication is no problem, if it ever was a problem; the DNA of the centromere replicates as does all other chromosomal DNA. Already we know that heterochromatin is late-replicating; heterochromatin can be at the ends of the chromosome as well as pericentromeric; not all chromosomes demonstrate heterochromatin in the vicinity of the centromere. So far as the localized centromere is concerned, centromeric DNA must be later replicating than any other kind, and there is no need to involve heterochromatin in its lateness. On the other hand heterochromatin may include many A-T sequences. Since replication of DNA occurs during interphase, and the presence of the kinetochore is not usually very apparent until prophase, the visible association of protein structures with the centromere region might not delay separation at all, since from early prophase the sister kinetochores are seen to be separate, and even at interphase (Earnshaw and Laemmli, 1983) they may be separate in G_2, although not in G_1. If some specific kinetochoric protein is already bound to CEN DNA at G_1, and during S, then perhaps part of the delay in later separation of sister kinetochores could be due to obstruction by such bound protein.

When the chromosomes are holokinetic, sister chromatids are not held together at any one region for longer than any other. It is often noticeable, in the alga *Spirogyra,* where the writer has seen hundreds of mitoses, that from their first alignment on the metaphase plate, sister chromatids are separate from end to end; sister chromatids are never distinguishable as such unless they are separated, with the single exception of anaphase II of meiosis where chromatids which might in part be sisters following terminalization of chiasmata are connected at their ends by half-chiasmata for a brief period (Godward, 1961). Even following fixation for the electron microscope, and the embedding schedules, this separation of sister chromatids is apparent at metaphase (Mughal and Godward, 1973), when it is not apparent in the angiosperm *Luzula nivea* (Bokhari and Godward, 1980), similar fixation and embedding schedules notwithstanding. There can be no doubt that fixation with alcohol-acetic acid, and with glutaraldehyde or osmic acid at 0°C will give different information about the chromosomes and much of classic light-microscope cytology is based on alcohol-acetic acid, which probably tells us basically more, by getting obstructive molecules out of the way or reducing their volume. From this point of view, it may be that early models of the centromere region (Lima-de-Faria, 1958) may assist in construction of a framework for those molecules.

E. Occurrence of the Diffuse Kinetochore

In the plant kingdom, holocentric chromosomes occur or have been reported in three monocotyledonous families: (1) Cyperaceae—*Eleocharis, Cyperus,* and *Carex;* (2) Juncaceae—*Luzula* and *Juncus* (probable); and (3) Liliaceae—*Chionographis.* In the algae, they have been reported in Chlorophyceae, Conjugales: *Spirogyra,* possibly other Zygnemaceae, and Desmidaceae. In the animal kingdom, they occur in larger taxonomic groups: Lepidoptera, Hemiptera, Homoptera, possibly *Ascaris* (*Parascaris*), and *Caenorhabditis;* in lower eukaryote, they occur in the ciliate *Nyctotherus.*

Not all these genera have been studied ultrastructurally. Among the plant genera *Luzula* has been the subject of many investigations. As reported by Braselton (1971) the chromosomes of *Luzula purpurea* are polycentric, the individual sites of spindle attachment being sunken into the surface of the chromosome (Fig. 14). In *Luzula nivea* however, continuous kinetochore extends along the length of the chromosome (Bokhari and Godward, 1980) but does not occupy its whole width; the kinetochore projects from the surface of the chromosome; it is much less electron-dense than the chromosome, and the microtubules can be seen passing through it, each carrying a little electron-dense collar near its base, to the chromatin below (Fig. 15).

In *Cyperus* (Bokhari 1979) and in *Luzula albida* (Lambert, 1980) (Fig. 16) the kinetochores are apparently in the same position as in *Luzula nivea* but are much less easy to see since they are only slightly less dense than the chromatin. The chromosomes of *Chionographis,* and the other monocotyledons, are not yet known ultrastructurally.

Discussion of *Luzula* is complicated by the tendency of chromosomes to adhere laterally when fixed for the electron microscope with glutaraldehyde (Bokhari and Godward, 1980).

In the algae, the chromosomes of *Spirogyra majuscula* (Mughal and Godward, 1973) carry a rather thin and only vaguely trilaminar kinetochore. In this account, the authors reported the chromosomes as polycentric, but it is now thought that the two ends of a sinuous chromosome were sectioned through the kinetochore, but that it was missed in the center where the chromosome curved away.

Nevertheless numerous studies with the ordinary light microscope would seem to indicate that some species of *Spirogyra* can have a localised centromere, in at least some of their chromosomes (Godward 1954, 1966) (Fig. 17) while some appear to have a centromeric region which is stronger in the middle and less so on either side, so that the chromosome has a curved leading region at anaphase,

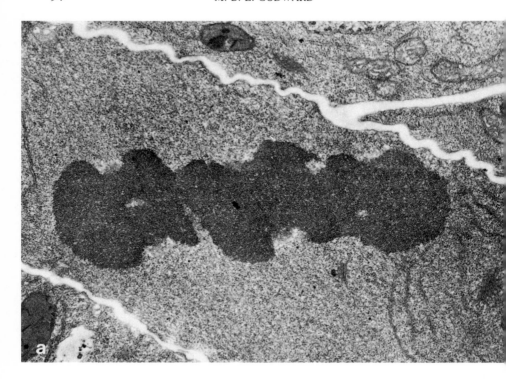

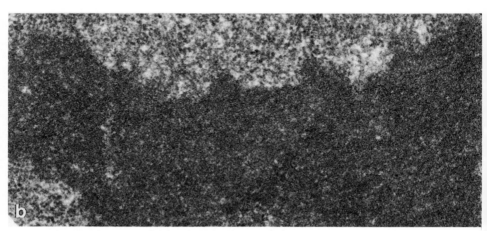

Fig. 14. (a) *Luzula purpurea*. Transversely sectioned metaphase, three chromosomes, each with sister kinetochores in depressions. Each kinetochore has an inner darker and outer light layer (Braselton, 1971). (b) *L. purpurea*. Oblique longitudinal section of early anaphase chromosome. Side-by-side depressions on the surface of the chromatid represent closely spaced kinetochores (Braselton, 1971).

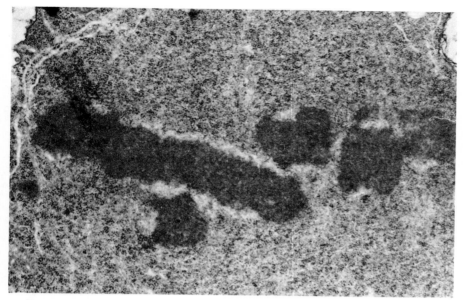

Fig. 15. *Luzula nivea* (from Bokhari and Godward, 1980). Diffuse kinetochore with prominent pale layer continuous along the longitudinally sectioned chromosome; microtubules pass through this layer, developing a small dark basal collar within it. Chromosomes cut transversely show that the kinetochore occupies a central strip only along the polar surfaces (×20,000).

instead of the V-shape of the localized centromere, or the horizontal stance of the diffuse or polycentric types.

The test of chromosome breakage by irradiation has been carried out on *Luzula* (La Cour, 1953; Bokhari and Godward, unpublished), *Spirogyra* (Godward, 1966) (Fig. 18), *Ascaris* (White, 1936; Bauer and Le Calvez, 1944), and *Caenorhabditis* (Albertson and Thomson, 1982). In all cases, chromosome fragments produced by irradiation can continue in mitosis and survive many cell divisions, and thus contain centromeric DNA. However, if a sufficiently high dose is given, there are eventually small fragments which are lost. It is possible that these may not be noncentromeric DNA intervening between centromeres (although they may be), but may contain damaged and inactivated centromere DNA, as might also be expected. Bridges and unequal daughter nuclei also occur. Following high doses that shatter the chromosomes, mitosis eventually stops.

There is some degree of assumption that whole groups—such as all Lepidoptera, Homoptera, and Hemiptera—have diffuse kinetochores, although if every additional example examined proves similar, this seems likely. There are certain features easily observable using ordinary cytological methods—the sinuous shape of the metaphase chromosome if it has length; the alignment with the

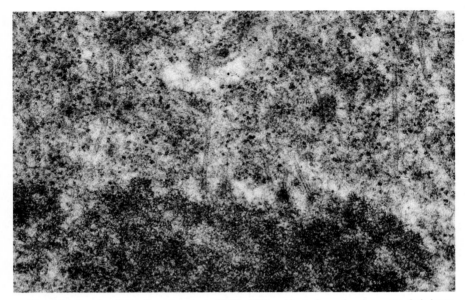

Fɪɢ. 16.   *Luzula albida* (from Lambert, 1980). Diffuse kinetochore showing outer dark layer composed of microtubule bases; pale layer not as prominent as in *L. nivea*.

equator of the spindle, maintained by the anaphase chromatids; the side-by-side position of members of a bivalent, with consequent equational first division at meiosis (Malheiros *et al.,* 1947; Hughes-Schrader, 1948; John and Lewis, 1965). When however the chromosomes are very small, or the nuclear divisions are in various ways aberrant, interpretation can be difficult.

Perhaps however one can say that whole taxonomic groupings in the animal kingdom may have diffuse kinetochores. In the plant kingdom, they are relatively restricted, two families of monocotyledons, one genus of a third family, and one group of green algae. In general, diffuse kinetochores are of random occurrence, scattered thinly over the plant and animal kingdoms, as if the diffuse condition could arise anytime by change and subsequently by evolutionary descent, to appear in related organisms now assembled into taxonomic groupings.

A few examples will be considered separately.

Rhodnius prolixus (Buck, 1967): The EM study by Buck (Figs. 1–8) indicates just the initiation of separation of sister chromatids at metaphase, and the apparently trilaminar kinetochore extending along the poleward surface (length) of the chromosomes. The author points out that few microtubule attachments to this superficial kinetochore layer were seen, and that many microtubules passed through the chromosomes completely. His illustrations also show (cf. section on the nuclear membrane) that nuclear membrane material is already forming in the

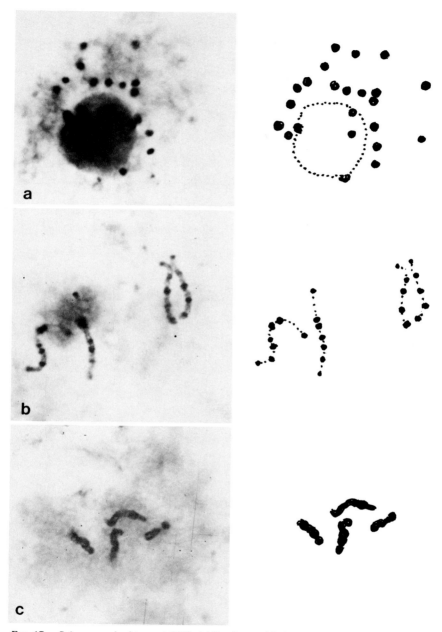

Fig. 17. *Spirogyra subechinata*, ×3000. (a) Prophase, with dot-like small chromosomes, showing faint linear connections in places. The adjacent diagram includes those visible at several focal levels and includes also the nucleolus. (b) Later prophase showing the dot-like structures now connected with fainter staining chromatin to form two pairs of chromosomes, one pair with five corresponding dots, and the other with seven and the nucleolar organizing region, clarified in the adjacent diagram. (c) Metaphase, showing four continuously darkly stained chromosomes.

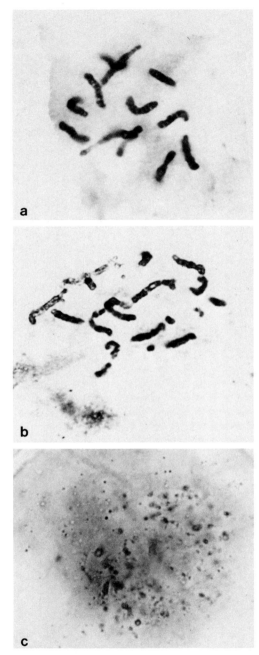

FIG. 18. *Spirogyra crassa:* (a) normal metaphase, 12 chromosomes; (b) after moderate irradiation; (c) after very high dose of irradiation; chromosomes shattered into fragments which still form a metaphase plate.

spindle and has almost completely surrounded each pair of anaphase chromatids—not yet fully separated. Data on the prophase are not available.

Ascaris lumbricoides (Goldstein, 1978): reconstruction of 17 serial sections through an autosomal bivalent at anaphase I showed a "diffuse and extensive association of the chromatin with microtubules" (Fig. 11) while a section through the metaphase I bivalent showed the chromosomes to be nearly covered over their surface with a layer typical of the trilaminar kinetochore, although the author describes it as a "pellicle." No "pellicles" were observed at anaphase II.

Since *Ascaris* has long been the favorite example of the polycentric chromosome, with numerous small chromosomes visible at prophase combining to form one or more long ones at metaphase, it must now appear that these are holocentric chromosomes like any others.

Nyctotherus ovalis (Eichenlaub-Ritter and Ruthmann, 1982) (Fig. 19): this is a ciliate, and the authors quote Devide and Geitler (1947) for the view that composite chromosomes are not uncommon in ciliate nuclei. Their Figs. 1–7 illustrate the presence of either 18 or 24 large chromosomes at metaphase (indicating the presence of two races and a basic chromosome number of 6 in the micronucleus). These larger chromosomes have been produced by the alignment side by side of many small chromosomes visible at prophase. Over the poleward surfaces (length) of the larger chromosomes a trilaminar kinetochore extends; but at anaphase the chromatids are again split into many small ones, each with a telomere composed of its share of the diffuse kinetochore. "While the single telokinetic chromosomes in *Nyctotherus* show a tendency to separate in early anaphase, they may still be held together by this mainly proteinaceous continuous outer kinetochore layer."

Both *Spirogyra crassa* (Godward, 1954) and *Chionographis japonica* (Tanaka, 1979), the first a green alga, the second a member of Liliaceae with metaphase chromosome numbers of 12 and 24, respectively, have in prophase numerous small "heteropycnotic segments" (Tanaka). In *Spirogyra crassa* the numbers of these ("stained blocks") were counted at different stages of the long-drawn-out prophase; they were progressively reduced from 356 to the metaphase number. The species *crassa* has some of the longest chromosomes in the genus; in another species the starting number was only 32, reduced to 24 at metaphase. There must be a lingering doubt as to whether all the 356 prophase units can really each carry a little kinetochore. Essentially, these organisms—*Ascaris, Nyctotherus, Spirogyra,* and *Chionographis* have in common the apparent junction of small chromosomal prophase units to give large holokinetic metaphase chromosomes. Only in *Nyctotherus* has the presence of a small individual kinetochore on each of the small chromosomal units been demonstrated with the electron microscope, unmistakable because it is trilaminar. Not all kinetochores are trilaminar and such a demonstration would be difficult if they were not. Often it is found that where investigations of metaphase and anaphase chromosomes are reported in detail, nothing is said about prophase, particularly early versus

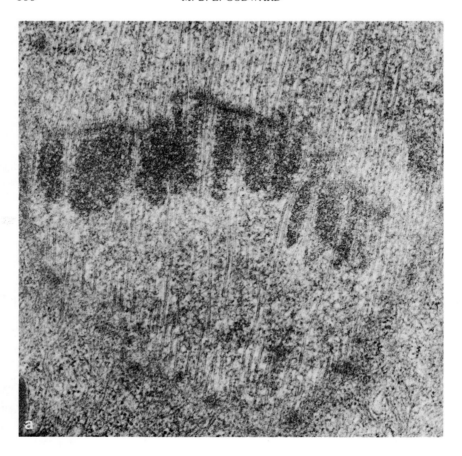

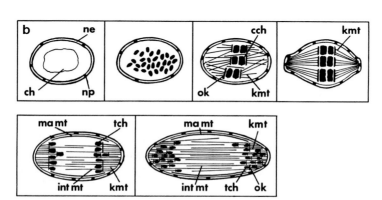

late prophase. It is fortunate that the vital evidence in *Nyctotherus* was visible at anaphase.

In *Caenorhabditis* (Albertson and Thomson, 1982) holokinetic chromosomes have recently been reported but there is no reference to prophase, although the authors do conclude, in agreement with Nordenskiöld (Nordenskiöld, 1951), that the holocentric chromosome might have evolved through fusion of small chromosomes.

Other questions might arise concerning the nature of such holokinetic chromosomes as have arisen in this way. In *Nyctotherus, Spirogyra,* and *Chionographis* the small units come together to form a definite number of chromosomes counted at metaphase. To what extent is there a tenuous connection between the small units at all times? It seems probable that there must be, or a constant chromosome number, and chromosome morphology (*Spirogyra subechinata,* Godward, 1966) would be impossible.

Another organism in which numerous small chromosomal units are visible in prophase, and only one large unit, a highly contracted metaphase plate, appears at metaphase, is the flagellate *Cryptomonas* (Hollande, 1942; Godward, 1966) (Fig. 4.2). While highly contracted metaphase plates are not uncommon in algae, large numbers of clearly defined discrete chromosomal units at prophase are not so common. Hollande described the metaphase structure as a "sammelchromosome," which would put *Cryptomonas* into a similar category to *Ascaris.* More recent accounts (Oakley and Bisalputra, 1977) have described the metaphase as a solid plate with tunnels for the microtubules. It could equally well be a number of closely contacting chromosomes. Electron-dense material at the base of a microtubule represents a kinetochore, but the distribution of microtubules over the metaphase plate is difficult to ascertain. The tendency of metaphase chromosomes, especially some holokinetic chromosomes, to adhere laterally when fixed for the electron microscope has recently been illustrated (Bokhari and Godward, 1980).

F. EVOLUTION OF THE DIFFUSE KINETOCHORE

It will be evident that if the localized kinetochore consists of DNA containing one or more CEN genes with associated sequences, plus the associated protein

---

FIG. 19. (a) *Nyctotherus ovalis* (from Eichenlaub-Ritter and Ruthman, 1982). Half spindle at early anaphase. Composite chromosomes (chromatids) dissociating into small telomeric elements. The trilaminar kinetochore still covers the poleward surface of the complex, extending across spaces between adjacent telomeric elements. ($\sim \times 60,000$) (b) Schematic drawing of mitosis in the micronucleus (see text for explanation). ne, nuclear envelope; np, nuclear pores; ch, chromatin; cch, composite chromosomes; ok, outer kinetochore layer; kmt, kinetochore microtubules; ma mt, manchette microtubules; int mt, interpolar microtubules; tch, telokinetic chromosomes.

molecules necessary for microtubule attachment, all that need be envisaged for the attainment of a diffuse kinetochore is the repetition of these sequences along the chromosome at intervals close enough for contact or near contact when the chromatin has sufficiently contracted, at late prophase or metaphase.

The region in the chromosome which becomes attached to the spindle—the "primary constriction" is, when examined under the visual light microscope, accompanied on either side by a region of "constitutive heterochromatin" which remains condensed when the "euchromatin" has expanded. As pointed out (Rieder, 1982), the existence of this pericentromeric heterochromatin has led, in some considerable degree, to confusion about the meaning of the terms centromere and kinetochore; he suggests that the term kinetochore be restricted to "the precise region on the chromosome that becomes attached to spindle microtubules"—presumably to the ultrastructurally distinct disc, strip, or ball-and-cup, to which microtubules can be seen attached, while the term centromere be used less precisely to include the primary construction plus pericentric heterochromatin. While the present writer has used the term kinetochore in the sense recommended above, now that the CEN genes have been isolated and sequenced, and shown to be effective when present on a piece of DNA too small to show any structure in the visual light microscope, the definition of centromere should surely be that region of the chromosome containing the CEN genes; its position commonly visible due to its behavior, but on occasion devoid of any specialised ultrastructure or any associated heterochromatin when seen in the light microscope.

If we now return to the problem of the diffuse kinetochore (the meaning of "centromere" having become somewhat theoretical), it is clear that, after chromatin contraction on passing from interphase to metaphase, there would be in most cases no way of determining, under the light microscope, how much noncentromeric DNA intervened between the repeated centromeric sections. There might be extensive stretches, greatly compressed; in which case the distinction between holocentric and polycentric would be only a matter of degree. It is hardly likely there would be no intervals between CEN sequences, since most of the chromosome has to code for varied transcriptions. If it is a question of the length of CEN DNA, another look at the very small kinetochores of *Polysphondylium, Physarum, Fusarium,* and other related organisms which attach only one microtubule, or only two or three, might suggest that the unit CEN gene, or only one or two contiguous copies, was present here. The usual kinetochore is considerably bigger and attaches several, perhaps 5 to 10, microtubules. Already then, there could be a stretch of repeats, in the localized kinetochore.

Under these circumstances, it seems quite reasonable that mutational events, breaks, translocations, duplications, and unions, involving kinetochore regions, could occur in any organism. The breaks and unions are widely known to be responsible for the appearance of both telomeric (Jones and Colden, 1968) and metacentric chromosomes in taxonomic groupings where they were not otherwise represented. The emergence of the diffuse kinetochore by such processes is obviously more complicated but nevertheless likely; witness the hybridization of *Luzula* species (Nordenskiold, 1951) where the small chromosomes of the species paired at meiosis with different parts of the long chromosomes of another species. (See Note Added in Proof.)

REFERENCES

Aist, J. R., and William, P. H. (1972). *J. Cell Biol.* **55,** 368–389.
Albertson, D. G., and Thomson, J. N. (1982). *Chromosoma* **86,** 409–428.
Aldrich, H. C. (1969). *Am. J. Bot.* **56,** 290–299.
Bauer, H. (1967). *Chromosoma* **22,** 249–281.
Bauer, H., and Le Calvez, J. (1944). *Chromosoma* **2,** 593–729.
Bloom, K. S., and Carbon, J. (1982). *Cell* **29,** 305–317.
Bokhari, F. S. (1979). Thesis, London University.
Bokhari, F. S., and Godward, M. B. E. (1980). *Chromosoma* **79,** 125–136.
Braselton, J. P. (1971). *Chromosoma* **36,** 89–99.
Brenner, S. L., Pepper, D., Berns, M. W., Tan, E., and Brinkley, B. R. (1981). *J. Cell. Biol.* **91,** 95–102.
Buck, R. C. (1967). *J. Ultrastruct. Res.* **18,** 489–501.
Burkholder, G. D. (1982). *Exp. Cell Res.* **142,** 485–488.
Byers, B., and Groetsch, L. (1975). *J. Bacteriol.* **124,** 511–523.
Darlington, C. D. (1936). *Cytologia* **7,** 242–247.
Darlington, C. D. (1937). "Recent Advances in Cytology," 2nd Ed. Churchill, London.
Dave, A. J. (1979). Thesis, London University.
Dave, A. J., and Godward, M. B. E. (1982). *J. Cell Sci.* **58,** 345–362.
Davies, D. R., and Plaskitt, A. (1970). *John Innes Annu. Rep.* p. 64.
de Brabander, M. *et al.* (1981). *J. Cell Biol.* **91,** 438–445.
Dubremetz, J. F. (1973). *J. Ultrastruct. Res.* **42,** 354.
Du Praw, E. J. (1966). *Nature (London)* **209,** 577.
Earnshaw, W. C., and Laemmli, U. K. (1983). *J. Cell Biol.* **96,** 84–93.
Eichenlaub-Ritter, A., and Ruthman, A. (1982). *Chromosoma* **84,** 701–716.
Fitzgerald-Hayes, C. and Carbon, J. (1982). *Cell* **29,** 235–244.
Gillott, M., and Triemer, R. E. (1978). *J. Cell Sci.* **31,** 25–36.
Girbardt, M. (1971). *J. Cell Sci.* **9,** 453–473.
Godward, M. B. E. (1954). *Ann. Bot.* **18,** 143–156.
Godward, M. B. E. (1961). *Heredity* **16,** 53–62.
Godward, M. B. E. (1966). *In* "The Chromosomes of the Algae," pp. 1–77. Arnold, London.
Goldstein, P. (1978). *Chromosoma* **66,** 59–69.
Guttes, S., Guttes, E., and Ellis, R. A. (1968). *J. Ultrastruct. Res.* **22,** 508–529.
Hadlaczky, G., Sumner, A. T., and Ross, A. (1981). *Chromosoma* **81,** 537–555.
Hakansson, A. (1968). *Heredita* **44,** 531–540.
Heath, I. B. (1980). *Int. Rev. Cytol.* **64,** 1–70.
Heath, I. B. (1981). *Int. Rev. Cytol.* **69,** 191–218.
Heath, I. B., and Greenwood, A. D. (1968). *J. Gen. Microbiol.* **53,** 287–289.
Heath, I. B., and Heath, M. C. (1976). *J. Cell Biol.* **70,** 592–607.

Hollande, A. (1942). *Arch. Zool. Exp. Gen.* **83**, 1–268.
Hollande, A., and Valentin, J. (1968). *C. R. Acad. Sci. Paris Ser. D.* **266**, 367–370.
Hughes-Schrader, S., and Ris, H. (1961). *Chromosoma* **12**, 327–350.
Ichida, A. A., and Fuller, M. S. (1968). *Mycologia* **60**, 141–155.
Jenkins, R. A. (1967). *J. Cell Biol.* **34**, 463–481.
John, B., and Lewis, K. R. (1965). *Protoplasmatologia* **VI**, 3–331.
Jones, K., and Colden, (1968). *Chromosoma* **24**, 135–157.
Kubai, D. F. (1973). *J. Cell Sci.* **13**, 511–552.
Kubai, D. F. (1975). *Int. Rev. Cytol* **43**, 167–227.
Kubai, D. F., and Ris, H. (1969). *J. Cell Biol.* **40**, 508–528.
La Cour, L. F. (1953). *Heredity (London)* **6**, 77–81.
Lafontaine, J. G., and Luck, B. T. (1980). *J. Ultrastruct.* **70**, 298–307.
Lambert, A. M. (1980). *Chromosoma* **76**, 295–308.
Lima-de-Faria, A. (1958). *Int. Rev. Cytol.* **7**, 123–156.
Lu, B. C. (1967). *Chromosoma* **22**, 210–226.
McCully, E. K., and Robinow, C. F. (1972). *J. Cell Sci.* **11**, 1–32.
McDonald, K. (1972). *J. Phycol.* **8**, 156–165.
McLaughlin, D. J. (1971). *J. Cell Biol.* **50**, 737–745.
Manton, I., Kowalkik, K. and Von Stosch, H. A. (1969). *J. Cell Sci.* **5**, 271–298.
Marchant, H. J., and Pickett-Heaps, J. D. (1970). *J. Phycol.* **6**, 189–215.
Marsden, M. P. F., and Laemmli, U. K. (1979). *Cell* **17**, 849–858.
Miller, O. L., and Bakken, A. M. (1971). *Acta Endocrinol. (Copenhagen)* **168**, 155–177.
Moens, P., and Rapport, E. (1971). *J. Cell Biol.* **50**, 344–361.
Mughal, S., and Godward, M. B. E. (1973). *Chromosoma* **44**, 213–229.
Nordenskiold, H. (1951). *Hereditas* **37**, 325–355.
Oakley, B. R., and Bisalputra, T. (1977). *Can. J. Bot.* **55**, 2789–2800.
Oakley, B. R., and Dodge, J. D. (1974). *J. Cell Biol.* **63**, 322–325.
Oakley, B. R., and Dodge, J. D. (1976a). *Cytobiologie* **17**, 35–46.
Oakley, B. R., and Dodge, J. D. (1976b). *Protoplasma* **88**, 241–254.
Oakley, B. R., and Heath, I. B. (1978). *J. Cell Sci.* **31**, 53–70.
Okada, T. A., and Comings, D. E. (1980). *Am. J. Hum. Genet.* **32**, 814.
Paulson, J. R., and Laemmli, U. K. (1977). *Cell* **12**, 817–828.
Pepper, D. A., and Brinkley, B. R. (1979). *J. Cell Biol.* **82**, 585–591.
Perkins, F. O. (1970). *J. Cell Sci.* **6**, 629–653.
Perkins, F. O. (1975). *J. Cell Sci.* **18**, 3277.
Peterson, J. B., and Ris, H. (1976). *J. Cell Sci.* **22**, 219–242.
Peterson, J. B., and Ris, H. (1976). *J. Cell Sci.* **22**, 219–242.
Pickett-Heaps, J. D., and Fowke, L. C. (1969). *Aust. J. Biol. Sci.* **22**, 857–894.
Pickett-Heaps, J. D., and Fowke, L. C. (1970). *Aust. J. Biol. Sci.* **23**, 71–92.
Pickett-Heaps, J. D., Tippit, D. H., and Andreozzi, J. A. (1978). *Biol. Cell.* **33**, 71–78, 79–84.
Pickett-Heaps, J. D., Tippit, D. H., and Porter, K. R. (1982). *Cell* **29**, 729–744.
Rattner, J. B., Krystal, G., and Hamkalo, B. A. (1978). *Chromosoma* **66**, 259–268.
Rattner, J. B., Goldsmith, M., and Hamkalo, B. A. (1980). *Chromosoma* **79**, 215–224.
Rieder, C. L. (1979). *J. Ultrastruct. Res.* **66**, 109–119.
Rieder, C. L. (1982). *Int. Rev. Cytol.* **79**, 1–57.
Ris, H., and Kubai, D. F. (1974). *J. Cell Biol.* **60**, 702.
Robinow, C. F., and Marak, J. (1966). *J. Cell Biol.* **29**, 129–151.
Roos, U. P. (1973). *Chromosoma* **41**, 195–220.
Roos, U. P. (1975). *J. Cell Biol.* **64**, 480–491.
Roos, U. P. (1977). *Cytobiologie* **16**, 82–90.

Roth, L. E. (1967). *J. Cell Biol.* **34,** 48–59.

Sakai, A., and Shigenaga, M. (1972). *Chromosoma* **37,** 101–116.

Scott, J., Bosco, C., Schornstein, K., and Thomas, J. (1980). *J. Phycol.* **16,** 507–524.

Shahanara, S. (1977). Thesis, London University.

Sharp, L. W. (1934). "Introduction to Cytology," 3rd. Ed.

Snyder, J., and McIntosh, J. R. (1975). *J. Cell Biol.* **67,** 744–760.

Tanaka, K. (1973). *J. Cell. Biol.* **57,** 220.

Tanaka, N. Y. (1979). *Cytologia* **44,** 935–949.

Tippit, D. H., and Pickett-Heaps, J. D. (1977). *J. Cell Biol.* **73,** 705.

Tippit, D. H., Pickett-Heaps, J. D., and Leslie, R. (1980). *J. Cell Biol.* **86,** 402–416.

Turner, F. R. (1980). *J. Cell Biol.* **37,** 370–393.

Utsumi, K. R. (1982). *Chromosoma* **86,** 683–702.

Vickerman, K., and Preston, T. M. (1970). *J. Cell Sci.* **6,** 365–383.

Wahl, H. A. (1940). *Am. J. Bot.* **27,** 458–470.

White, M. J. D. (1936). *Nature (London)* **137,** 183.

Zickler, D. (1970). *Chromosoma* **30,** 287–304.

NOTE ADDED IN PROOF.   The delayed "division" of the localized centromere, until Anaphase II of meiosis, is not of course a feature of the diffuse kinetochore nor could it be of the more or less polycentric condition.

INTERNATIONAL REVIEW OF CYTOLOGY, VOL. 94

The Evolved Chromosomes of Higher Plants

G. P. Chapman

*Department of Biological Sciences, Wye College, University of London,
Wye, Ashford, Kent, England*

I. Introduction

Currently, one of the most beckoning prospects in plant biology is that of being able to study isolated functional chromosomes *in vitro*. Moreover, in recent years concepts have changed about the chromosomes' receptiveness both to exogenous DNA and their relationship to other DNA-containing organelles. This article, therefore, is in two parts: the first attempts to redescribe the chromosome taking account of new results with some necessary reference to studies in animal cytology, and the second considers the very recently developed techniques for chromosome isolation and their role in further research.

II. Changing Chromosome Concepts

A. The Nuclear Genome

The chromosome is both an organelle and a macromolecule and those occurring in vascular land plants are recognizably, the outcome of evolutionary events. The familiar mitotic stages there should overshadow neither the range of nuclear form and function in lower eukaryotes nor the varied patterns of chromosome replication in angiosperms giving polyteny, diplochromosomes, polyploid nuclei, and multinucleate cells, summarized by Brown and Dyer (1972). Frisch and Nagl (1979) for example have provided for one species, *Scilla decidua,* a detailed assay of nuclear types present throughout the plant.

Grant (1963) argued that the basic number of primitive angiosperms was $n =$ 7–9, an approach elaborated for example by Ehrendorfer *et al.* (1968) to explain evolution among supposedly more primitive angiosperms. Stebbins (1971) suggested that metacentric chromosomes preceded acrocentrics (although Robertsonian centromere fusion would interrupt the trend on this assumption). Evolution here leading to the array of ploidies and karyotypes in present day angiosperms meant mutation and rearrangement exclusive to the chromosome, a concept now discarded. Earlier, Stebbins (1950) had attributed lapsed homology to "cryptic structural change" an idea nowadays increasingly accessible to molecular techniques. Hutchinson *et al.* (1980) for *Aegilops* × *Secale* hybrids used a repetitive sequence that could be *in situ* probed at meiosis to locate *Secale* chromosomes. Using a specific RNA probe, there workers found preferential pairing of arms without ribosomal gene clusters.

Evidence for interorganelle exchange of DNA among chromosomes, mitochondria, and chloroplasts as Ellis (1982) remarked diminishes the separate identity of each genome and gene exchange between pro- and eukaryotes together with progress in chromatin studies requires therefore a reappraisal of chromosome form and function. And again, if the original eukaryote linear chromosome were from ring chromosome prokaryote ancestors, the acquisition of kinetochores, plural replicons, and nucleosomes creates an area of study— recognizable but hardly explored.

B. LINEARIZATION

If a ring chromosome were opened (perhaps at the origin of replication), to remain linear a mechanism is required to maintain and replicate free ends. Cavalier-Smith (1974) proposed the evolution of a foldback palindromic telomere from a bacterial restriction mechanism so as to remove the difficulty of replication of the 3′ end of a molecule of linear DNA. Bateman (1975) simplified this idea, the essentials of which are that the self-palindrome is normal in G_1, G_2, and M but is lost in S. The endonuclease nicking the palindrome was assumed telomere specific. Bateman (1975) also suggested that the telomeres of one genome all carried the same palindrome. Szostak and Blackburn (1982) cloned presumed yeast telomeres using a Tetrahymena linear plasmid, one conclusion being that such sequences were very highly conserved.

Dinoflagellates are recognizably eukaryote. Are their circular chromosomes similar to those from which the linear eukaryote arose? Their chromatin is nonnucleosomal and incorporates heavy metals, some thymidine is substituted by 5-hydroxymethyl uracil (Dodge, this volume), and they are curiously "quasi-polytene." Even assuming their derivation from an earlier unineme, they seem, at present, unlikely candidates to originate typical linear eukaryote chromosomes. Cavalier-Smith (1981) argued that ancestral dinoflagellates lost genes for histones but survived because of an unusual mitosis regarded as a preadaptation for histone loss.

Hollande (1974) using *Oxyrrhis, Peridinium, Solenodinium,* and *Syndinium* found that both histone content and chromosome structure varied, and perhaps it is among these and related organisms that one must search for the beginnings of eukaryote chromosome phylogeny.

Although in *Saccharomyces cerevisiae* the telomere is a terminal palindrome, the chromosome end in higher plants is even with light microscopy, a distinctive structure now known to contain DNA sequences varying among related organisms and divisible into more and less highly conserved parts. At the EM level fiber ends are seldom seen and if folded within the chromosome body (not to be confused with a terminal palindrome), the unineme end would not coincide with the visible chromosome end.

C. THE KINETOCHORE

Electron microscopy clearly shows for primitive organisms, an intimate involvement of the kinetochore with the nuclear membrane; Kubai (1975) and Dodge and Godward (this volume) provide examples. In higher plants, the nuclear membrane has disintegrated prior to spindle formation and the kinetochore is a permanent chromosome feature. Tanaka and Tanaka (1974) demonstrated for *Polygonatum* species that kinetochores were stainable throughout the cell cycle and after division, clustered on opposite sides of daughter nuclei "in the nuclear periphery adjacent to the nuclear membrane." Hadlaczky (this volume) shows how in protein-depleted plant chromosomes the centromeres (and occasionally telomeres) remain as the ultimate detectable structural entities, a situation seen in similarly treated mammal chromosomes and emphasizing the kinetochore–chromosome association.

Ris and Witt (1981) using cultured mouse cells (with a trilaminar kinetochore) reversibly dispersed and reassembled the outer lamina by changes in the ionic strength of applied KCl and suggested the outer zone to be a special arrangement of chromatin fiber spun out from the chromosome body and lacking a nucleosome structure. As in yeast, quoted here by Godward, the microtubules were considered directly attached to the chromatin fibers.

No similar study is yet available for higher plants although in the diffuse kinetochore of *Luzula nivea,* Bokhari and Godward (1980) detected microtubule attachment through the kinetochore pad to the chromosome body. The direct structural connection (if any) between the kinetochore and a nonhistone protein core is understood for neither plants nor animals.

Unlike yeast, many microtubules penetrate higher plant kinetochores implying amplification of the appropriate gene complex. *Haemanthus katherinae* has 50 to 150 microtubules per kinetochore (Bajer, 1968). Figure 1 is of a barley centromere minus the disc but offering a large area for microtubule attachment. In animals, Moens (1978), for a grasshopper species *Neopodismopsis,* demonstrated in Robertsonian fusions a doubling of microtobule number at the kinetochore. Paired kinetochore discs on well-separated chromatid centromeres are commonly seen. Almost certainly, interdigitated chromatin fiber preserves the paired centromere arrangement until the microtubule mechanism parts them.

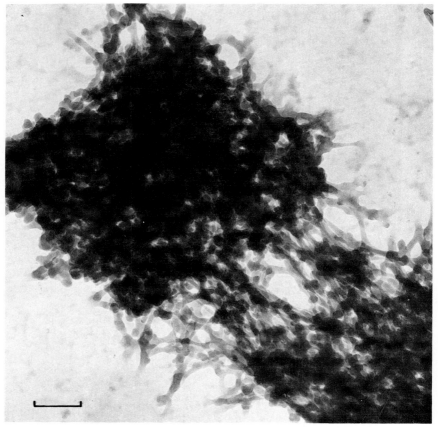

FIG. 1.   Transmission EM detail of an acrocentric *Hordeum vulgare* chromosome showing short arm and centromere region. Bar = 0.2 μm. Photo courtesy H. Allam.

Orientation of the kinetochore relative to the centromere is precise. Unfixed HeLa chromosomes at metaphase when supplied with tubulin can become sites for microtubule assembly unless face downward to the slide and inaccessible to the tubulin (Telzer *et al.*, 1975). Assuming the kinetochore is on the "obverse" face of the centromere, it is this and not the "reverse" that would align to the metaphase plate.

Holmquist and Dancis (1979) considered telomeres and kinetochores in relation to Robertsonian fusion. Their telocentric chromosome is one where a kinetochore plate overlaps a telomere. When one considers the possible unfolding of a terminal region the convoluted fibrillar structure of a metaphase chromosome might still have the kinetochore organizer region and the telomere proper separated by a substantial length of chromatin, a matter yet to be resolved by electron microscopy at which level a true telocentric chromosome (that is immediate juxtaposition of kinetochore and telomere) may not exist.

D. The Chromosome Body

The halo of DNA (Hadlaczky, this volume) is a dispersed macromolecule that on integration with other chemicals gives an organelle with a hierarchy of ordered structures reckoned to regulate gene action and differentiation. Structurally it can be approached either by ultrathin sectioning (Bennett *et al.*, 1983), or by isolation of whole chromosomes. Such whole chromosomes fixed either before or after isolation and with varying degrees of dispersion can be prepared for electron microscopy. Chapman and Cooke (1983) developed a technique to isolate *Vicia faba* chromosomes for light, transmission, and scanning EM and since shown applicable to *Hordeum* and *Linum* species. These authors using in *V. faba* the translocation homozygotes of the EF karyotype where each chromosome in the set is recognizable developed by Michaelis and Rieger (1971) were found to preserve chromosome identity.

Tarawali (unpublished), combining the EM evidence such as that in Fig. 2 and the known DNA values proposed for *V. faba* a metaphase chromatin assembly as follows. Assuming a c value of 14 pg for six chromatids (Mayer and Fox, 1973; Bennett and Smith, 1976) and equal proportions of G-C and A-T base pairs this is 4.25×10^6 μm DNA. Packing into nucleosomes and then into a 25-nm fiber gave an estimated 50-fold length contraction or 8.5×10^4 μm in six chromatids. For simplicity, one normal untranslocated acrocentric chromosome and lacking a secondary constriction and representing 12.5% of the total DNA was chosen for consideration. Such a chromatid, 5 μm long, would contain 1.0625×10^4 μm of 25-nm fiber requiring a 2125-fold packing. This could be achieved by a fiber running back and forth for short distances and integrating domains of looping parallel to and at right angles to the chromatid long axis. This concept of "packing domains" could be harmonized with both zones of banding and differ-

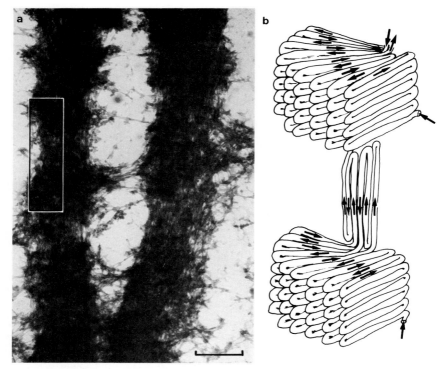

FIG. 2.   (a) Portion of a chromatid pair from a *Vicia faba* chromosome at early metaphase. Bar =
0.5 μm. (b) Diagrammatic interpretation of boxed portion. Photo courtesy S. A. Tarawali.

ential decontraction and perhaps stabilized by nonhistone protein. Interestingly,
Sumner (1982) did not relate banding to any version of the looped unineme
concept.

Recalling the previous section dealing with placement of the kinetochore,
ideally perhaps the metaphase chromosome is, for any given type within a
species, a precisely ordered bundle where left and right chromatids are exact
copies (not mirror images) with characteristic obverse and reverse topographies
and interphase decondensation would be equally precise, a view reconsidered
later in the discussion of isolated chromosomes.

E. Eu- and Heterochromatin, Content and Function

Differential staining can identify "heterochromatin" which, in molecular
terms, can be shown rich in repetitive DNA. Two recent studies combining these
approaches extend our understanding of heterochromatin.

Maize has heterochromatin proximal to the centromere and at the N.O.R. Heterochromatin can additionally be present as "knobs" and in supernumerary B chromosomes. Each replicate at different times (Pryor *et al.*, 1980). Peacock *et al.* (1981) have demonstrated specific for knob heterochromatin a 185 base pair repeating sequence. What is of special interest is that a large knob (K10) on chromosome 10 has about 10^6 copies of this sequence. However, this knob additionally can induce neocentric activity in the other knobs, a property presumably attributable to some other, as yet unknown, sequence.

Within the Triticeae, there is evidence of base pair sequence amplification and rearrangement, the extent of which approximates with taxonomic relationship (for a review see Flavell, 1980). Bennett (1977) proposed that in Triticale, a wheat × rye hybrid, grain shrivelling was attributable to late replication of rye telomeric heterochromatin resulting in unresolved anaphase bridges in coenocytic endosperm. Telomeric heterochromatin consists of three base pair repeat families in *Secale cereale*, 480, 610, and 630 that separate it from *S. sylvestre* (Bedbrook *et al.*, 1980). Since the 480 and 610 repeats are absent from wheat it was possible to test whether Triticales lacking such heterochromatin showed improved grain fill. Bennett and Gustafson (1982) and Gustafson and Bennett (1982) concluded that where recognizable blocks of telomeric heterochromatin were omitted improvement could result in one or more of the following: nuclear stability in early endosperm development, kernel weight, and yield. The effects were not wholly beneficial since one block of heterochromatin, when lost, was associated with a decrease in fertility.

In both the maize and triticale examples the presence or absence of satellite DNA is significant. Part of the explanation involving the location of single or few-copy gene sites remains difficult, although in animals this has been achieved using salivary gland chromosomes, hence the recently renewed interest in plant polyteny.

F. PLANT POLYTENY

Drosophila polytene studies are now highly developed. Hill and Stollar (1983), utilizing micromanipulation of individual chromosomes into a buffer considered to preserve higher order structure, presented evidence for the adoption of z-DNA formation following liberation from torsional stress. Wu and Davidson (1981) combining *in situ* hybridization with colloidal gold marking demonstrated, at the EM level, location of single gene sites to quote two examples. Polytene nuclei occur at crucial points in the plant life cycle but because of inaccessibility, have generated few parallels with animal studies. Brady (1973) for suspensor cells of *Phaseolus* species described a diffuse replicating phase alternating with a more compact one and stated that while euchromatin replication occurred first, there was complete rather than partial replication of the chromosome. Barlow (1975) reported polyteny, curiously, in giant hair cells of *Bryonia dioica* anthers. Nagl (1982) has reviewed DNA endoreplication among plants generally.

An area of interest economically is the role of polytene antipodal nuclei in cereal grain fill. In Gramineae, antipodal cells vary from about four in *Pennisetum americanum* (Rao *et al.*, 1983) with perhaps a quarter of the cells becoming trinucleate eventually to *Sasa paniculata* with about 300 uninucleate antipodals (Yamaura, 1933). Approximate numbers for common cereals include *Secale cereale* 15, *Triticum aestivum* 25, and *Hordeum vulgare* 50 (Bennett *et al.*, 1973). According to these authors, antipodal development in wheat cv. "Chinese Spring" coincides with the 2 nucleate pollen stage with 20 to 30 cells formed by second pollen division. DNA values by endoreplication reach up to 200 times the haploid value 4 days after pollination. Such nuclei are, in size, larger than a wheat pollen grain. Differences in appearance were noted—a resting stage (DNA synthesis?) and prophase—recalling Brady (1973). Morrison (1955) observing disintegration of antipodal nuclei by about 9 days concluded they had no leading role in endosperm nutrition.

These huge cereal nuclei pose two related problems. The first is their precise role whether "secretory" or "sacrificial" in regard to early endosperm development and how, if it all, shrivelled grain in cereal hybrids can be attributed to antipodal malfunction. Heslop Harrison (1972) grouped antipodals on the basis of time of arrest of DNA and chromonemal replication relative to cell division giving either polyteny, endopolyploidy, multinucleate cells, or a syncitium. If one phase were normal for a species, hybridization might shift the antipodal nuclei to another, disruptively. This aspect remains to be investigated. The second problem is whether with advancing microtechnique, the antipodal tissue can be utilized for *in vitro* and ultrastructural studies.

Brink and Cooper (1944) compared antipodals in *Hordeum jubatum* selfs with those produced when crossed with *Secale cereale* where in the latter case their development was slow and restricted and these authors concluded, fatally impaired early endosperm development. Kaltsikes *et al.* (1975) for Triticale found that for lines with earliest degeneration of antipodals, these showed fastest *early* endosperm growth, most production of chromosome aberrants, and, subsequently, the most shrunken grain.

No published *in vitro* studies of cereal antipodal tissue are known to the author. Among experimental systems perhaps the most promising is that for *Pisum sativum* where polytene chromosomes in cotyledon culture were described by Marks and Davies (1979). Subsequently, Davies and Cullis (1982) have shown that polytene nuclei can be readily accumulated in defined medium using the variety "Greenshaft" and several others. These authors used cRNA as a probe for *in situ* hybridization to presumed sites of nucleolar organizers. Amplification to 32c or 64c DNA was reported. These authors considered feasible,

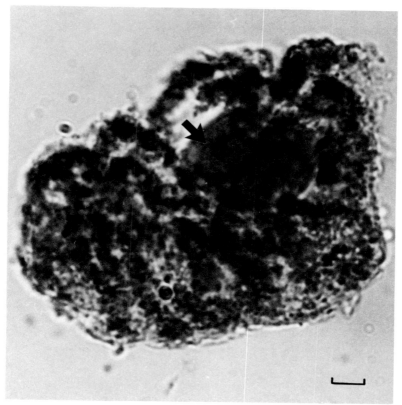

FIG. 3. Isolated fixed polytene nucleus of *Pisum sativum*, toluidine blue stained. Nucleolus arrowed. Bar = 10 μm. Photo courtesy of P. M. Allington.

extension of the technique to *Arachis*, *Glycine*, *Phaseolus*, and *Vicia* where in the last case Cionini *et al.* (1978) appear, from their illustrations, to have obtained polyteny in callus culture.

Work in this laboratory has shown that these polytene nuclei can be readily extracted from cultured cotyledon tissue and handled with a micro-manipulator (see Figs. 3 and 4). Preparations derived from such nuclei show under transmission EM skeins of parallel chromatin fibers (see Fig. 5). No exact equivalent of banding or puffing has yet been detected.

G. CHROMOSOME REARRANGEMENT DURING DIFFERENTIATION

Gross chromosome variation can occur in culture, presumably adaptively. A recent review is that of Bayliss (1980). Modification of the karyotype within

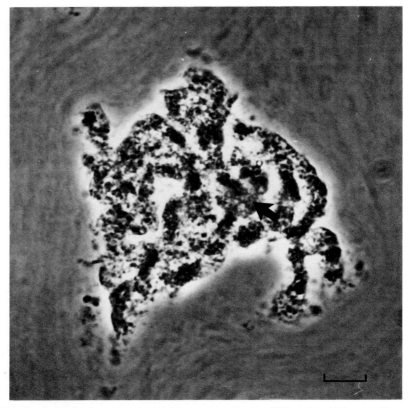

FIG. 4.   Isolated polytene nucleus of *Pisum sativum* fixed in 45% acetic acid showing the speck-led zones of varied density in the prophase chromosomes. Nucleolus arrowed. Bar = 10 μm. Photo courtesy of P. M. Allington.

differentiating plants can occur as with ''megachromosomes'' with the abnormal situation in *Nicotiana* involving specific hybrids and involving amplification of either eu- or heterochromatin (Gerstel and Burns, 1967, 1976). Endoreplication of entire genomes occurs in some but not all species during differentiation (Evans and van t'Hoff, 1975). Nagl (1976) found endoreplication more likely where the 2c value was low. However, the case for *selective* amplification or rearrangement of the karyotype as a strictly normal feature of differentiation is more difficult to substantiate. Either, therefore, this kind of chromosome change is below the level of optical resolution or it does not occur. However, Pearson *et al.* (1974) gave evidence for a higher proportion of satellite DNA in meristematic cells than in those that have endopolyploidized during tissue development. Al-though the role of controlling elements in reprogramming development in maize

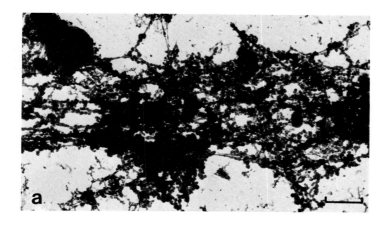

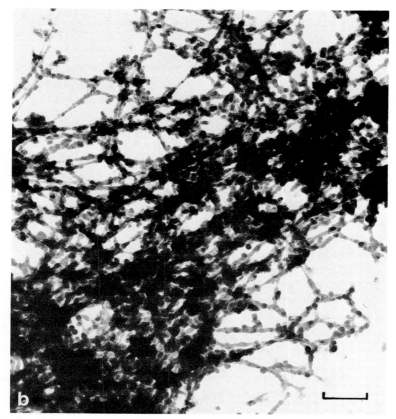

Fig. 5. (a) Part of a single polytene chromosome of *Pisum sativum* viewed by TEM. Zones of varied density again evident. Bar = 1.0 μm. (b) Enlarged portion of a single *Pisum sativum* polytene chromosome. Note in both a and b overall parallel strandedness with varied lateral strandedness. Bar = 0.5 μm. Photo courtesy P. M. Allington.

grain is well documented, the importance of such elements as fundamental to gene regulation in development has not been widely accepted (Federov, 1983).

H. INTERORGANELLE DNA EXCHANGE

Ward *et al.* (1981) drew attention to the relative constancy of nuclear genome size coupled with an 8-fold increase in mitochrondrial genomes among related species of Cucurbitaceae and contrasted this to that for two *Vicia* species where despite large differences in nuclear DNA content, mitochondrial DNA values were similar. They concluded that genome size in the two organelles varied independently.

Since the nuclear genome is both the largest and that indispensible to all functional cells and assuming a symbiont origin for chloroplasts and mitochrondria, phylogenetically, the nucleus must have been more a recipient than a donor of DNA. Ellis (1983) reviewed the case of *Cyanophora paradoxa* where ribulose biphosphate carboxylase/oxygenase is coded in one genome. This enzyme in higher plant chloroplasts is coded partly in the chloroplast but the small regulatory units are coded from the nucleus thus implying transfer of coding sequences.

Ontogenetically, within the lifetime of the organism, small-scale transfer bestowing some temporary advantage might, theoretically, take place in either direction between any two of the three DNA organelles.

According to Heslop-Harrison (1980), at pollen division mitochondria migrate to the generative nucleus and plastids do so in Gramineae but not Orchidaceae for example. Thereafter both types of organelles if present are probably completely removed from male nuclei on passage through the synergid. If the male contribution were exclusively nuclear then, in the zygote, nuclear enrichment by the "chondriome" (mitochondria plus chloroplast genomes) would be maternal in origin but in reverse the chondriome could be enriched from either or both parents. In beginning the next generation either acquired sequence changes being merely ontogenetic, are corrected perhaps as part of a meiotic "clean up" or, if uncorrected, become heritable, if only in the short term. Possibly significant here is the extreme simplification of plastid structure coincident with pachytene described by Dickinson and Heslop Harrison (1977), Dickinson and Potter (1978), and Medina *et al.* (1981). Stern and Lonsdale (1982) have shown homology between maize mitochondrial and chloroplast DNA for a 12 kilobase region and detected for cytoplasmic male sterile variants, a deletion in this region of the mitochondrial genome which viewed in this way, survives any supposed organelle reorganization during female meiosis.

Any transfer of DNA from the chondriome to the nucleus is from a non-nucleosomal to a nucleosomal organization. J. Ellis (personal communication) points out on the basis of DNA injection into *Xenopus* nuclei that additional

DNA is likely to form nucleosomes since the process is not sequence specific. Since chloroplast and nuclear promotors are not the same, there would have to be sequence changes to adapt to the new transcriptional environment. Although chloroplast DNA is nonnucleosomal, it is apparently elaborately condensed with protein (Briat *et al.*, 1982).

I. Gene Insertion from beyond the Cell

The addition of plasmid DNA from *Agrobacterium tumefaciens* both diminishes exogenous hormone dependency of plant tissue *in vitro* and, if appropriately modified, can survive meiosis and demonstrate Mendelian segregation (Otten *et al.*, 1981). Robins *et al.* (1981) transformed rat liver cells using direct uptake of DNA coding for a variant human growth homone. A range of transformed lines was produced, the chief points of interest here being that from 5 to 100 copies of the gene were inserted in different lines, the site of integration differed, gene copies tended to insert in tandem at one site, in some lines the point of insertion was associated with detectable chromosome rearrangement, and where transforming DNA was inserted near an rDNA site, this latter was apparently inactivated, In yeast, multiple copies of an inserted gene are also known to integrate in tandem (Struhl, 1983).

Pandey (1978) and Caligari *et al.* (1981) with increasing irradiation of pollen found that the contribution of the male genotype to the progeny could be diminished yielding "matromorphs." Among more likely explanations, either a whole but largely incapacitated male genome was handed on or the male chromatin was "pulverized" and fragments only of it, incorporated into the subsequently doubled female genome. For *Nicotiana rustica* a tetraploid, if the endosperm were hexaploid, presumably a male nucleus was available and by implication to the zygote, too and if not then presumably a "less than hexaploid" endosperm was able to sustain an unusual embryo but as yet the cytology of endosperm here has not apparently been examined. Pandey (1980) for the hybrid *N. bonariensis* ($2n = 18$) × *N. glauca* ($2n = 24$) found $2n = 18$ rather than $2n = 21$. Since the matromorph was fertile and the sexual hybrid sterile, doubling of the female genome seems the more likely explanation. Even here it is not clear what happens in the endosperm.

The precise mechanism of DNA (or chromatin) insertion for *Nicotiana* matromorphs is not understood. Certain genes, notably those for the S self-incompatibility system and flower color genes, appear preferentially inserted. Pandey (1978) argued that if heterochromatin rich in repetitive DNA repairs nonspecifically, those genes might owe their frequent inclusion to their association with the appropriate heterochromatin. The matter remains unresolved.

Clearly, within the cell environment all manner of chromosome change is possible. Removal of chromosomes from the cell for *in vitro* studies is now increasingly feasible for both normal and modified chromosomes and the implications are next considered.

III. Isolation and Manipulation of Functional Chromosomes

A. ANIMAL CYTOLOGY

Since the studies of Wray and Stubblefield (1970) isolation of essentially functional mammal metaphase chromosomes has become routine and led to such developments as chromosome mediated gene transfer (CMGT) shown by McBride and Ozer (1973) to permit gene control of synthesis of hypoxanthine phosphoribosyltransferase to be removed from a hamster to a mouse genome in cell culture. For a review of this and related developments in genetic analysis, see Ruddle (1981). Sillar and Young (1981) and Davies *et al.* (1981), using human chromosome suspensions treated with ethidium bromide, showed that these can be fractionated by flow cytometry and as techniques improve the prospects for rapid accumulation of mass samples of a single chromosome type.

B. PLANT STUDIES

Only very recently have unfixed presumably functional, plant chromosomes been isolated but the reasons for doing so are significant. A corpus of technique is already available from animal studies. There is the prospect of handling pure samples of intact organelles either singly or as mass samples or in predetermined combinations under defined conditions to study both form and function. And there is, too, the prospect of devising a technique for precise transfer of parts of one genome into another.

Published work and the genera utilized include the following. Malmberg and Griesbach (1980), *Hemerocallis* and *Lilium* (meiosis), *Lycopersicon,* and *Nicotiana* (mitosis); Griesbach *et al.* (1982), those genera previously mentioned and *Allium, Pisum, Vicia,* and *Zea* (mitosis); Szabados *et al.* (1981), *Petroselinum* and *Triticum* (mitosis); Griesbach *et al.* (1982), those previously mentioned and *Alium, Pisum, Vicia,* and *Zea* (mitosis); Hadlaczky *et al.* (1983), *Triticum* and *Papaver* (mitosis); Matthews (1983), *Daucus* (mitosis).

These pioneer studies identify a common goal, uptake of foreign chromosomes by protoplasts for eventual incorporation of part of the donor genome into that of the host. For chromosome isolation, although differing in detail, these studies share a similar approach outlined as follows.

1. Acquisition of rapidly dividing tissue, either root tip or suspension culture

where the mitotic index is if possible raised further using fluorodeoxyuridine or hydroxyurea.

2. Spindle inhibition and chromosome contraction using colchicine.

3. Conversion enzymically, of cells to protoplasts.

4. Syringing and centrifugation of protoplasts to release chromosomes into a carefully formulated buffer.

This list of genera includes many familiar to cytologists and is sufficiently diverse to suggest that the technique could be widely applicable, but presently, mitotic indices seldom exceed 30% compared with about 95% in *Homo sapiens* synchronized cell cultures.

C. Utilization of Isolated Chromosomes: Future Priorities

Given the availability of such chromosomes this will surely prompt studies similar to those with isolated mitochrondria and chloroplasts. How "intact," with present techniques, are isolated chromosomes? This must relate partly to isolation procedure and partly to the buffer within which chromosomes are collected and retained.

Hall (1972) for isolated chloroplasts classified them according to the degree of damage to which scheme Lilley *et al.* (1975) proposed amendments as details accumulated. The present author proposes here the following tests to assess the intactness of isolated chromosomes.

1. *Unfixed*

1. Both chromatids are at any stage equally condensed or decondensed.

2. Under appropriate conditions centromeres of isolated metaphase chromosomes will elaborate microtubules.

3. With slight changes of buffer, the chromosomes can be reversibly compacted and dispersed to provide a simple test of responsiveness other than at the centromere.

4. Under sterile conditions, the features indicated in 1–3 will not deteriorate with time.

2. *Fixed*

5. When prepared for electron microscopy fiber diameter and pattern of distribution are identical for both chromatids.

Buffers used vary widely (see Table I). Griesbach *et al.* (1982) stress that buffers must be at or near neutral pH to preserve protein composition and include metal chelators to bind cofactors that would otherwise allow nuclease activity. This latter, they claim, is further reduced by colchicine, low temperature, high ionic strength, and the presence of polyamines which induce chromosome con-

TABLE I
Chromosome Buffers[a]

Reference	Sucrose (mM)	Spermine (mM)	Spermidine (mM)	EDTA (mM)	EGTA (mM)	KCl (mM)	NaCl (mM)	Mercaptoethanol (mM)	Tris–HCl or HEPES or PIPES (mM)	Hexylene glycol (mM)	Other ingredients (mM)
Animal (small sample only)											
Blumenthal et al. (1979)	340	0.2	0.5	2	0.5	80	20	14	15 Tris–HCl pH 7.2	500	
Sedat and Manuelidis (1978)	110	0.15	0.5	0.5	0.1	80	15	15	15 Tris–HCl pH 7.4		
Lewis and Laemmli (1982) Mg-Hex buffer	17%							0.1 PMSF + 1% thiadiglycol	0.1 PIPES pH 6.5	1	1 $MgCl_2$
Lewis and Laemmli (1982)							10	0.1 PMSF + 1% thiadiglycol	10 Tris–HCl pH 7.4		
Aqueous buffer			0.25	2							
Aqueous buffer				K-EDTA		2					
Polyamine buffer		2[b]	0.8[b]								
Gooderham and Jeppesen (1983)				0.5 added in some cases		120	20		10 Tris–HCl pH 8		
Plant											
Cheah and Osbourne (1977)	300	0.15	0.5	—	—	10	15	5 DTT	15 Tris–HCl		3% PVP, 0.5% colchicine, 0.025% BSA, pH 6.5
Malmberg and Griesbach (1980)									50 MES		5 $CaCl_2$
Griesbach et al. (1982)	300	0.5		1	—	80	20	DTT 15	HEPES 15 pH 7.0	500	
Hadlaczky et al. (1983)								1 PMSF		1%	100 glycine pH 8.4–8.6 (Ca(OH)$_2$); 1% isopropyl alcohol
Matthews (1983)									20 Tris–HCl pH 7.0		1 $MgCl_2$ 1 $CaCl_2$ 1 $ZnCl_2$

[a]The table includes a sample of contrasted animal chromosome buffers and those utilized for plant studies. Detergent additions such as Triton, Digitonin, and Brij have been omitted. My thanks are due to Mrs. S. A. Tarawali who prepared the original table and offered much helpful comment.

[b]Changed during isolation.

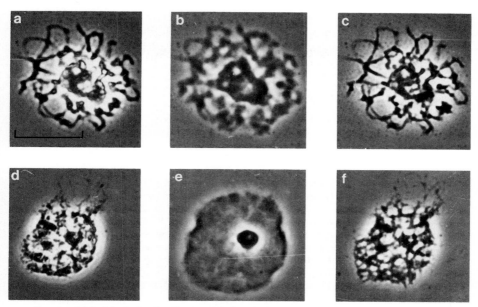

FIG. 6. *Vicia faba* isolated unfixed nuclei viewed via phase contrast to show effects of varying buffer conditions. (a) and (d) Hepes buffer alone; (b) and (e) Hepes buffer plus 0.215 *M* sodium chloride; (c) and (f) NaCl removed using Hepes buffer. Bar = 20 μm. Photo courtesy S. A. Tarawali.

traction. Even a simple change in buffer composition can have a dramatic effect. Figure 6 shows two unfixed *Vicia faba* prophase nuclei[1] dispersed and recompacted by the effects of NaCl addition and removal. Each returns to its previous stage rather than to earlier or later stages. The process is repeatable up to a dozen times after which chromosomes collapse (Tarawali, 1983).

Tests 1–4 above could provide a basis for the efficacy of isolation procedures and the appropriateness of buffers and test 5 extends these jointly with fixing and mounting techniques, to electron microscopy. As regards future priorities, apart from a better understanding of buffers, it is clearly useful to raise the mitotic index further and improve the efficiency of chromosome retrieval. Therefore, presumably, attention will turn to the ultrastructure of these macromolecules in various states of controlled unravelling in relation to gene activity. And it is to be hoped meiotic and polytene chromosomes, those with diffuse centromeres, and the peculiar chromosomes of Dinoflagellates will be included.

For plant cells, it is worth remarking that so far, the most stable incorporation of endogenous DNA has required its substantial fragmentation whether by bacte-

[1]This process is not identical with decondensation and condensation, hence the alternative terms.

rial plasmid transfer or by matromorph induction. A functionally intact remotely alien chromosome must be disruptive and unstable in its host protoplasts and, logically, the emphasis must now be on functional chromosomes as DNA or chromatin fragment receptors.

REFERENCES

Bajer, A. (1968). *Chromosoma* **25**, 249–281.
Barlow, P. W. (1975). *Protoplasma* **83**, 339–349.
Bateman, A. J. (1975). *Nature (London)* **253**, 379–380.
Bayliss, M. W. (1980). *Int. Rev. Cytol. Suppl.* **11A**, 113–144.
Bedbrook, J. R., Jones, J., O'Dell, M., Thompson, R. D., and Flavell, R. B. (1980). *Cell* **19**, 545–560.
Bennett, M. D. (1977). *Heredity* **39**, 411–419.
Bennett, M. D., and Gustafson, J. P. (1982). *Can. J. Genet. Cytol.* **24**, 93–100.
Bennett, M. D., and Smith, J. B. (1976). *Philos. Trans. R. Soc. London Ser. B* **274**, 227–275.
Bennett, M. D., Rao, M. K., Smith, J. B., and Baylis, M. W. (1973). *Philos. Trans. R. Soc. London Ser. B* **266**, 39–81.
Bennett, M. D., Heslop Harrison, J. S., Smith, J. B., and Ward, J. P. (1983). *J. Cell Sci.* **63**, 173–179.
Blumenthal, A. B., Dieden, J. D., Kapp, L. N., and Sedat, J. W. (1979). *J. Cell Biol.* **81**, 255–259.
Bokhari, F. S., and Godward, M. B. G. (1980). *Chromosoma* **79**, 125–136.
Brady, T. (1973). *Caryologia* **25**, (Suppl.), 233–259.
Briat, J. F., Gigot, C., Laulhere, J. P., and Mache, R. (1982). *Plant Physiol.* **69**, 1205–1211.
Brink, R. A., and Cooper, D. C. (1944). *Genetics* **29**, 391–406.
Brown, R., and Dyer, A. F. (1972). "Plant Physiology: A Treatise" (F. C. Stewart, ed.), Vol. V, pp. 49–90. Academic Press, New York.
Caligari, P. D. S., Ingram, N. R., and Jinks, J. L. (1981). *Heredity* **47**, 17–26.
Cavalier-Smith, T. (1974). *Nature (London)* **250**, 467–470.
Cavalier-Smith, T. (1981). *Soc. Gen. Microbiol. Symp.* **32**, 33–83.
Chapman, G. P., and Cooke, S. A. (1983). *Protoplasma* **116**, 198–200.
Cheah, K. S. E., and Osborne, D. J. (1977). *Biochem J.* **163**, 141–144.
Cionini, P. G., Bennici, A., and D'Amato, F. (1978). *Protoplasma* **96**, 101–112.
Davies, D. R., and Cullis, C. A. (1982). *Plant Mol. Biol.* **1**, 301–304.
Davies, K. E., Young, B. D., Elles, R. G., Hill, M., and Williamson, R. (1981). *Nature (London)* **293**, 374–376.
Dickinson, H. G., and Heslop Harrison, J. (1977). *Philos. Trans. R. Soc. London Ser. B* **277**, 327–342.
Dickinson, H. G., and Potter, U. (1978). *J. Cell Sci.* **29**, 147–169.
Ehrendorfer, F., Krendl, F., Habeler, E., and Sauer, W. (1968). *Taxon* **17**, 337–353.
Ellis, J. (1982). *Nature (London)* **299**, 678.
Ellis, J. (1983). *Nature (London)* **304**, 308–309.
Evans, L. S., and vant'Hof, J. (1975). *Am. J. Bot.* **62**, 1060–1064.
Federov, N. V. (1983). *In* "Mobile Genetic Elements" (J. A. Shapiro, ed.), pp. 1–63. Academic Press, New York.
Flavell, R. B. (1980). *Annu. Rev. Plant Physiol.* **31**, 569–596.
Frisch, B., and Nagl, W. (1979). *Plant Syst. Evol.* **131**, 261–276.

Gerstel, D. U., and Burns, J. A. (1967). *Genetics* **56**, 483–502.
Gerstel, D. U., and Burns, J. A. (1976). *Genetica* **46**, 139–153.
Gooderham, K., and Jeppesen, P. (1983). *Exp. Cell Res.* **144**, 1–14.
Grant, V. (1963). "The Origin of Adaptations." Columbia Univ. Press. New York.
Griesbach, R. J., Malmberg, R. L., and Carlson, B. S. (1982). *Plant Sci. Lett.* **24**, 55–60.
Gustafson, J. P., and Bennett, M. D. (1982). *Can. J. Genet. Cytol.* **24**, 93–100.
Hadlaczky, G., Bisztray, G. Y., Praznovszky, T., and Dudits, D. (1983). *Planta* **157**, 278–285.
Hall, D. O. (1972). *Nature (London) New Biol.* **235**, 125–126.
Heslop Harrison, J. (1972). *In* "VIC Plant Physiology: A Treatise" (F. C. Steward, ed.), pp. 133–289. Academic Press, New York.
Heslop Harrison, J. (1980). *Philos. Trans. R. Soc. London Ser. B* **292**, 588.
Hill, R. J., and Stoller, B. D. (1983). *Nature (London)* **305**, 338–340.
Hollande, A. (1974). *Protistologia* **10**, 413–43451.
Holmquist, G. P., and Dancis, B. (1979). *Proc. Natl. Acad. Sci. U.S.A.* **76**, 4566–4570.
Hutchinson, J., Chapman, V., and Miller, T. E. (1980). *Heredity* **45**, 235–254.
Kaltsikes, P. J., Roupakias, D. G., and Thomas, J. B. (1975). *Can. J. Bot.* **53**, 2050–2067.
Kubai, D. F. (1975). *Int. Rev. Cytol.* **43**, 167–228.
Lewis, C. D., and Laemmli, U. K. (1982). *Cell* **29**, 171–181.
Lilley, R. McC., Fitzgerald, M. P., Fienitz, K. G., and Walker, D. A. (1975). *New Phytol.* **75**, 1–10.
McBride, O. W., and Ozer, H. (1973). *Proc. Natl. Acad. Sci. U.S.A.* **70**, 1258–1262.
Malmberg, R. L., and Griesbech, R. J. (1980). *Plant Sci. Lett.* **17**, 141–147.
Marks, G. E., and Davies, D. R. (1979). *Protoplasma* **101**, 73–80.
Matthews, B. F. (1983). *Plant Sci. Lett.* **31**, 165–172.
Mayer, E. P., and Fox, D. R. (1973). *Nature (London) New Biol.* **245**, 170–172.
Medina, F. J., Riseuno, M. C., and Rodriguez-Garcia, M. I. (1981). *Planta* **151**, 215–225.
Michaelis, A., and Rieger, R. (1971). *Chromosoma* **35**, 1–8.
Moens, P. B. (1978). *Chromosoma* **67**, 41–54.
Morrison, J. W. (1955). *Can. J. Bot.* **33**, 168–176.
Mullinger, A. M., and Johnson, R. T. (1979). *J. Cell Sci.* **38**, 369–389.
Nagl, W. (1976). *Nature (London)* **261**, 614–615.
Nagl, W. (1982). *In* "Cell Growth" (C. Nicolini, ed.), pp. 619–651. Plenum, New York.
Otten, L. De Grieve, H., Hernalsteens, J. P., and van Montagu, M. (1981). *Mol. Gen. Genet.* **183**, 209–213.
Pandey, K. K. (1978). *Genetica* **49**, 56–69.
Pandey, K. K. (1980). *Heredity* **45**, 15–29.
Peacock, W. J., Dennis, E. S., Rhoades, M. M., and Pryor, A. J. (1981). *Proc. Natl. Acad. Sci. U.S.A.* **78**, 4490–4494.
Pearson, G. G., Timmis, J. N., and Ingle, J. (1974). *Chromosoma* **45**, 281–294.
Pryor, A., Faulkner, K., Rhoades, M. M., and Peacock, W. J. (1980). *Proc. Natl. Acad. Sci. U.S.A.* **77**, 6705–6709.
Rao, M. K., Kumari, K. A., and Grace, J. R. (1983). *Bot. Gaz.* **144**, 201–206.
Ris, H., and Witt, P. L. (1981). *Chromosoma* **82**, 153–170.
Robins, D. M., Ripley, S., Henderson, A. S., and Axel, R. (1981). *Cell* **23**, 29–39.
Ruddle, F. H. (1982). *Nature (London)* **294**, 115–120.
Sedat, J., and Manuelidis, L. (1978). *Cold Spring Harbor Symp. Quant Biol.* **XLII**, 331–350.
Sillar, R., and Young,B. D. (1981). *J. Histochem. Cytochem.* **29**, 74–75.
Stebbins, G. R. (1950). "Variation and Evolution in Plants." Columbia Univ. Press, New York.
Stebbins, G. L. (1971). "Chromosomal Evolution in Higher Plants." Arnold, London.
Stern, D. B., and Lonsdale, D. M. (1981). *Nature (London)* **299**, 698–702.

Struhl, K. (1983). *Nature (London)* **305,** 391–397.
Sumner, A. T. (1982). *Cancer Genet. Cytogenet.* **6,** 59–87.
Szabados, L., and Dudits, D. (1980). *Exp. Cell Res.* **127,** 442–446.
Szabados, L., Hadlaczky, G., and Dudits, D. (1981). *Planta* **151,** 141–145.
Szostak, J. W., and Blackburn, E. H. (1982). *Cell* **29,** 245–255.
Tanaka, N., and Tanaka, N. (1974). *Cytologia* **42,** 765–775.
Tarawali, S. A. (1983). Ph.D. thesis, University of London.
Telzer, B. R., Moses, M. J., and Rosenbaum, J. L. (1975). *Proc. Natl. Acad. Sci. U.S.A.* **72,** 4023–4027.
Ward, B. L., Anderson, R. S., and Bendich, A. J. (1981). *Cell* **25,** 793–803.
Wray, W., and Stubblefield, E. (1970). *Expl. Cell Res.* **59,** 469–478.
Wu, M., and Davidson, N. (1981). *Proc. Natl. Acad. Sci. U.S.A.* **78,** 7059–7063.
Yamaura, A. (1933). *Bot. Mag.* **47,** 551–555.

NOTE ADDED IN PROOF. On exogenous gene transfer: see DeWet, J. M. J. *et al.,* ''The Experimental Manipulation of Ovule Tissues.'' Pitmans, London (in press). On flow cytometric sorting of plant chromosomes: see de Laat, A. M. M., and Blaas, J. (1984). *Theor. Appl. Genet.* **67,** 463–467.

INTERNATIONAL REVIEW OF CYTOLOGY, VOL. 94

Structure and Biochemistry of the Sertoli Cell

Donald J. Tindall, David R. Rowley, Latta Murthy,
Larry I. Lipshultz, and Ching H. Chang

Department of Cell Biology, Baylor College of Medicine, Houston, Texas

I. Introduction

Spermatogenesis is a highly complex process which is regulated directly and indirectly by the pituitary hormones, luteinizing hormone (LH) and follicle stimulating hormone (FSH). The primary function of LH is to stimulate the production of testosterone within the Leydig cells of the testis. Both testosterone and FSH act on the seminiferous epithelium to influence spermatogenesis. FSH initiates spermatogenesis, while testosterone maintains spermatogenesis. The Sertoli cell is in intimate contact with the germ cells during all phases of spermatogenesis, which includes mitotic proliferation, meiotic division, and differentiation or spermiogenesis. The juxtaposition of Sertoli cells and germ cells has led many investigators to believe a priori that Sertoli cells play an important regulatory role in spermatogenesis. More recent advances in understanding the biochemistry of the Sertoli cell have substantiated this belief and have led to the theory that the Sertoli cell is involved in mediating the hormonal regulation of spermatogenesis.

In this article, we will review three aspects of the Sertoli cell: its structural and biochemical components, and the hormonal regulation of the Sertoli cell by FSH and testosterone.

127

II. Structural Components

The close morphological relationship between Sertoli cells and the developing germ cells offers the Sertoli cell a unique position whithin the testis. Sertoli cell processes extended laterally, surrounding germ cells and provide a support network such that each germ cell is in contact with a number of adjacent Sertoli cells (Nicander, 1963; Nagano, 1966; Fawcett, 1975). Therefore, in order to understand the role of the Sertoli cell in spermatogenesis a clear understanding of Sertoli cell morphology is important.

A. The Blood–Testis Barrier

The testis is divided into several compartments. These compartments are important in separating not only cell types but also hormonal and biochemical components. The interstitial compartment is composed of Leydig cells, macrophages, blood vessels, and lymphatic vessels. The seminiferous tubular compartment is composed of a peritubular layer of myoid cells, Sertoli cells, and germ cells. Additionally, the seminiferous tubular compartment is divided into an adluminal compartment and a basal compartment by the blood–testis barrier (Fawcett, 1975). The functional significance of the blood–testis barrier was first demonstrated by studies using electron opaque materials. These studies demonstrated that substances of a certain molecular size will pass through the testis and appear in the lymph from the testis but not in the lumen of the rete testis (Waites and Setchell, 1969). The basal compartment is accessible to substances from blood via the extracellular spaces. However, because of the occluding nature of the blood–testis barrier, these substances are prevented from reaching the adluminal compartment. These observations suggested that the blood–testis barrier was not located in the walls of the interstitial capillaries, but rather was located on or within the seminiferous tubule. Thus, the extracellular spaces in the adluminal compartment would presumably contain only those substances which are produced or transported by the Sertoli cell. The blood–testis barrier, therefore, may play an important role in the regulation of spermatogenesis. For example, the mitotically proliferating germ cells (i.e., spermatogonia and preleptotine spermatocytes) residing in the adluminal compartment are separated from those germ cells in the basal compartment, which are undergoing meiosis and spermiogenesis and may require a different milieu for their cellular processes.

The blood–testis barrier is composed of two components (Dym and Fawcett, 1970; Fawcett et al., 1970). One component is the peritubular layer of myoid cells. Junctional complexes between adjacent myoid cells will exclude certain electron opaque materials throughout most of the seminiferous tubule. In some regions of the tubules, however, the tracers do penetrate the myoid cell layer due to the lack of the junctional complexes in these regions. The second component of the blood–testis barrier is composed of tight junctional complexes between adjacent Sertoli cells (Connell, 1978; Nagano and Suzuki, 1978). These junctional complexes are located in the basal lateral cell membranes of adjacent Sertoli cells and establish the adluminal and basal compartments of the extracellular space.

Junctional complexes develop between Sertoli cells at between 16 and 19 days of age in the rat (Vitale *et al.*, 1973). At this time the sex cords become canalized and the blood–testis barrier is established. Thenceforth, interstially injected electron opaque tracers are effectively prevented from reaching the tubule lumen (Dym and Fawcett, 1970; Fawcett, 1970; Vitale *et al.*, 1973). In order to study the temporal correlation of these events, a biochemical marker was needed to determine when the sex cords became fully canalized, and the blood–testis barrier became functional. Since androgen binding protein (ABP) was shown to be produced in the rat testis, secreted into the lumen of the seminiferous tubules, and carried to the epididymis via the efferent ducts, this appeared to be an appropriate marker. Therefore ABP was measured during postnatal development in both testis and epididymis of rats to determine whether development of a blood–testis barrier and formation of a continuous lumen from testis to epididymis correlated with entry of ABP into the caput epididymis (Tindall *et al.*, 1975). ABP binding activity was found in the testis as early as 14 days postnatally (0.2 pmol/mg protein) at which no blood–testis barrier was observed by the peroxidase perfusion technique. Previous findings had demonstrated a close correlation of blood–testis barrier development and lumen formation. Indeed, ABP was not detected in the caput epididymis until 18–20 days of age (1 pmol/mg protein) at which time both blood–testis barrier formation and lumen development were complete. These findings suggested that entry of ABP into the caput epididymis could be used as an index of blood–testis barrier formation and lumen development (Tindall *et al.*, 1975). Although hypophysectomy is known to prevent the rat testis from producing ABP, this treatment does not affect the ability of the junctional complexes to exclude lanthanum, nor does it change their fine structural details (Hagenas *et al.*, 1978). Therefore, it appears that the occluding function of the blood–testis barrier is not regulated by gonadotropins.

In higher primates and humans the blood–testis barrier appears to be composed primarily of the intraepithelial component, that is the Sertoli cell junctional complexes (Fawcett, 1975). Unlike in rodents, the peritubular myoid cells in humans may not play a role in a functional blood–testis barrier. Indeed, the Sertoli cell tight junctions served as a useful marker in transmission electron micrographs in identifying these cells, which were isolated from human testes and grown in culture (Lipschultz *et al.*, 1982). These complexes retained their morphological features throughout the isolation procedure, which involved mechanical separation of the tissue, sequential trypsin and collagenase enzyme digestion, and disruption of tubules by passage through a wire mesh grid. Thus these interesting morphological features were indeed tight junctions and were very difficult to disrupt.

Other cytoplasmic structures exist in close proximity to the junctional complexes forming the blood–testis barrier. In the cytoplasm opposite these junctions, subsurface cisternae of endoplasmic reticulum, and bundles of microfilaments are adjacent to opposed Sertoli cell membranes (Fawcett, 1975). During the stages of the spermatogenic cycle, tubulobular complexes exist between adjacent Sertoli cells in the region of the blood–testis barrier as well as between spermatids and Sertoli cells (Russell, 1979, 1981). These complexes consist of an evagination of one cell into an adjacent cell (2–4 μm in length) with the resulting structure having a tubular region at the evagination site with a bulbous dilation at the terminal end (Russell, 1981). These tubulobular complexes undergo a pattern of dynamic changes during the spermatogenic cycle by increasing during stages II–V of the cycle and undergoing regression during stages VI–VII. During stages VII–XIV-I the few remaining complexes do not demonstrate the typical bulbous swelling. Between spermatids and Sertoli cells, the tubulobular complexes undergo regression, and newly formed complexes appear, suggesting that there is a cyclic turnover of these complexes. At the time of sperm release all the complexes are absorbed at the top of the spermatid head, and new complexes form on the lateral aspect.

There are also desmosome-like junctions (Gravis, 1979) and gap junctions (McGinely et al., 1979) between Sertoli cells and spermatogonia and spermatocytes. It appears likely that the desmosome junction has an adherens-like capacity to connect adjacent cells to one another. The gap junctions may serve a more physiological role by providing a mode or route for cell-to-cell communication or "signaling," although the term "cellular communication" is poorly defined, and the role or roles of the gap junctions are not yet known. However, junctional complexes and other intracellular specializations may be potentially important for the process of spermatogenesis and the synchronization of the seminiferous epithelium.

B. GENERAL FINE STRUCTURAL CHARACTERISTICS

A number of morphological features of the human Sertoli cell distinguish it from other cell types within the testis. In order to identify Sertoli cells isolated from human testis and grown as primary cultures in vitro, it was necessary to demonstrate that these cells had morphological features similar to Sertoli cells in vivo (Lipschultz et al., 1982; Tindall et al., 1983). Testes were obtained from estrogen treated men undergoing male-to-female sex reassignment surgery. The germ cell population in these testes had been completely depleted by the estrogen treatment and was therefore an excellent source of Sertoli cells. Interestingly, these human testes were strikingly similar to the Sertoli cell-enriched rat testes which had been irradiated *in utero* and in which there was a complete absence of germ cells (Tindall and Means, 1976). The isolation procedure involved first mincing the tissue and extensive washing with Hanks' balanced salt solution. After allowing the free tubules to settle, the tubular fragments were treated with 0.1% collagenase (Class IV, 165 U/mg) and 0.1% trypsin in a shaker bath at 1500 oscillations/minute for 15 minutes at 37°C. The supernatant was removed and passed through a wire mesh screen (mesh size 0.25 mm). The undigested tissue was subjected to further enzymatic digestion. Any tissue remaining after these treatments was found to consist of peritubular myoid cells and Leydig cells and was discarded. The supernatant fraction was treated with 5% fetal calf serum for trypsin neutralization. This preparation consisted primarily of Sertoli cell aggregates. These aggregates were dispersed by passage through a 17-gauge needle. The cells were concentrated by centrifugation, and the pellet was washed with Eagles MEM containing D-valine, pen-strep, and 5% fetal calf serum. Cells were plated at a concentration of 10^7 viable cells per 25 cm^2 in 5 ml of MEM and incubated at 37°C in 5% CO_2 and 95% O_2. Attachment of these cells was visualized usually within 2–3 hours of plating and a complete monolayer was identified by 24 hours.

One of the most outstanding features of these Sertoli cells was the prominent nucleolus and associated perinucleolar bodies. The nucleolus of Sertoli cells is centrally located and probably functions as a site for the transcription of ribosomal RNA. Perinuclear bodies are in juxtaposition to the nucleus and at opposite sides (Fawcett, 1975; Krimer, 1977; Coleman and Stockhert, 1978). The morphological characteristics of the pernucleolar bodies have led investigators to interpret them as being condensed chromatin (Fawcett, 1975; Coleman and Stockhert, 1978). The function of these perinculeolar bodies is not understood. They may function as sites for posttranscriptional modifications of ribosomal RNA or for associating mRNA with ribosomal RNA prior to transport from the nucleus; however, this remains a speculation. Interestingly, many of the human Sertoli cells in culture retained their irregularly shaped nucleus, a distinguishing feature of most Sertoli cells *in vivo,* where the axis of the ovoid nucleus is oriented in parallel to the cell axis and is generally located in the basal region of the cell (Nagano, 1966; Krimer, 1977). Nuclear pores are present within the irregular foldings of the nuclear envelope, and the chromatin is highly euchromatic, lacking peripheral clumps of heterochromatin (Fawcett and Burgos, 1956; Fawcett, 1975; Krimer, 1977).

Another distinguishing feature of human Sertoli cell cultures is the conspicuous aggregation of smooth surfaced endoplasmic reticulum (SER), often arranged concentrically in multiple layers around lipid droplets (Lipshutz *et al.,* 1982; Tindall *et al.,* 1983). These arrays have been observed to surround pleomorphic lipofuscin pigment and secondary lysosomes (Bawa, 1963; Nagano, 1966; Fawcett, 1975). The concentric arrangement of SER around a spherical lipid inclusion is suggestive of a steroid producing function of the Sertoli cell (Christensen, 1965; Christensen and Fawcett, 1966). Indeed, Sertoli cells possess the capacity to convert testosterone to estradiol, and this process is stimulated by FSH (Dorrington et al., 1978; Welsh and Wiebe, 1978). However, the physiological significance of this steroid in the testis is not clear since the Leydig cell also produces significant quantities of estradiol. Also of significance is the concentric arrays of SER around the Sertoli cell membranes adjacent to the acrosome region of the spermatids (Clermont et al., 1980). Whether this arrangement is responsible for steroid metabolism of some kind or is possibly a component of the phagocytosis and enzyme digestive processes is not yet known. The human Sertoli cells demonstrated this concentric arrangement of agranular reticulum (Lipshultz et al., 1982; Tindall et al., 1983). In the center, a body of electron-dense material and several vesicles containing amorphous material was often observed. The arrangement of SER around secondary lysosomes and pleomorphic lipofuscin pigment suggests that they may play a role in the enzymatic digestive process of the Sertoli cell. Indeed, the Sertoli cell was first described as a phagocytitic cell, and the phagocytitic nature of the Sertoli cell may be quite important for the normal process of spermatogenesis. Sertoli cells phagocytose degenerating germ cells that die normally during the spermatogenic process (Black, 1971; Fawcett, 1975). Moreover, when mature spermatozoa are released, the residual cytoplasm left behind are phagocytosed by the Sertoli cells (Black, 1971; Fawcett, 1975). Lysosomes not only digest substances phagocytosed by the Sertoli cell, but also are involved in the natural autolytic process of degenerating cytoplasmic organelles.

An important function of the Sertoli cell is the active secretion of a variety of proteins. Cytoskeletal elements of a cell provide the necessary support and transport mechanisms for secretory vesicles and products, thereby aiding in the process of secretion. Morphological features of the human Sertoli cell suggested that these cells were active in protein synthesis and secretion (Lipshultz et al., 1982; Tindall et al., 1983). Abundant Golgi apparatus was evident. Traditional views of the Golgi apparatus dictate that its chief function is to modify and package the synthesized products from the endoplasmic reticulum. These products are then eventually packaged into membrane bounded vesicles such as secretory granules and lysosomes. The Golgi complex in the Sertoli cell in vivo has been described by several investigators as composed of four to six flattened parallel cisternae with associated small vesicles (Nagano, 1966; Flickinger, 1967). The profiles of Golgi are also observed more frequently in the basal zone near the nucleus. A three-dimensional configuration of the entire Golgi apparatus of the rat Sertoli cell has been constructed using thick sections and high-voltage electron microscopy (Rambourg et al., 1979). These studies have demonstrated that the Golgi is indeed a very complex organelle in the Sertoli cell.

Because of the abundance of proteins produced and secreted by the Sertoli cell, two morphological observations are surprising. One is that the content of rough endoplasmic reticulum (RER), whose chief function is protein synthesis, is sparse in comparison to the content of agranular SER in the Sertoli cell (Nagano, 1966; Flickinger, 1967; Steinberger *et al.*, 1975). Tubular profiles of RER are observed more often than vesicular profiles and are localized predominantly in the basal region of the cell (Fawcett, 1975). The tubular profiles are observed sometimes in parallel arrangements, but are usually more randomly organized.

The other observation is that very few "secretory vesicles" are seen in the Sertoli cell both *in vivo* or *in vitro*. This leads one to suggest that the proteins secreted by the Sertoli cell are needed immediately by the surrounding germ cells. One exception to this is the secretion of androgen binding protein (ABP). A polyclonal antibody to ABP was raised in rabbits and used to localize ABP in rat Sertoli cells in culture, using an immunocytochemical technique. A preferential pattern of fluorescent granules was found in the perinuclear region in what appeared to be vesicular granules. Fluorescent material was also observed throughout the cytoplasm. However, not all of the Sertoli cells contained equivalent amounts of immunoreactive material. Sertoli cells treated with preimmune serum or ABP antiserum preabsorbed with purified ABP demonstrated only background immunofluorescence. Additionally, a ductus deferens tumor cell line (DDT_1: Norris and Kohler, 1976), which contains androgen receptor showed only background fluorescence, indicating no cross-reactivity to the androgen receptor. Finally, a hepatoma cell line (Hep G-2: Khan *et al.*, 1981), which produces and secretes human sex steroid binding protein (SBP), showed only background fluorescence, indicating no cross-reactivity with SBP. These results are similar to the findings of Kierszenbaum *et al.* (1980). Whether or not these discrete regions of ABP immunoreactivity are indeed secretory vesicles or simply localization of ABP in the Golgi region remains to be elucidated by EM analysis.

The human Sertoli cells also contained membrane bounded dense bodies which were evenly distributed throughout the cytoplasm (Lipshultz *et al.*, 1982; Tindall *et al.*, 1983). Such dense bodies *in vivo* are variable in size and include both primary lysosomes and probably secretory vesicles (Fawcett, 1975; Nagano and Suzuki, 1978). In addition to membrane bound granules, the Sertoli cell contained inclusions of lipofuscin pigment inclusions which were more pleomorphic in shape (Fawcett, 1956; Bawa, 1963; Nagano, 1966).

The human Sertoli cell cytoskeleton consisted of microfilaments, intermediate filaments, and microtubules (Lipshultz *et al.*, 1982; Tindall *et al.*, 1983). An elaborate array of microtubules, which extended to the periphery of the cell, were observed by indirect immunofluorescence microscopy. Microfilaments are distributed throughout the Sertoli cell cytoplasm and in some cases existed as bundles oriented parallel to the cell axis (Nagano, 1966; Fawcett, 1975). More specifically microfilaments are also associated with peripheral cytoplasm in close proximity to both Sertoli–germ cell and Sertoli–Sertoli cell junctional regions. The microfilaments in the Sertoli cell are actin-like in nature (Toyama *et al.*, 1979). Crystalloid structures composed of bundles of microfilaments, 150 Å in diameter, are present in the human Sertoli cell cytoplasm and have been termed Charcot-Botcher crystals (Lubarsch, 1896). The Sertoli cell also contains an abundance of the 7–11 nm intermediate filaments, which form a meshwork of variable arrangements throughout the cytoplasm, but are most prevalent in the juxtanuclear and apical regions (Franke *et al.*, 1979). These intermediate filaments are of the viminten type (Franke *et al.*, 1979). These findings are of interest because a primitive type of intermediate filament is found in non-muscle cells of "mesenchymal origin."

The human Sertoli cells *in vivo* contain abundant mitochondria, which are slender and had transversely arranged, incomplete cristae (Bawa, 1963; Nagano, 1966; Fawcett, 1975). The mitochondria usually have a moderately dense matrix and are oriented parallel to the cell axis of the supranuclear cytoplasmic region and demonstrate a more random arrangement in the basal region. Generally there is a greater concentration of mitochondria in cell types requiring a high utilization of ATP, and the distribution of mitochondria tends to be increased in areas of the cytoplasm with high energy requirement. The more structured orientation of mitochondria in the supranuclear region may be the result of different energy requirements in this region of the cell.

III. Biochemical Components

A. ANDROGEN BINDING PROTEIN

It is now well established that androgen binding protein (ABP) is produced and secreted by the Sertoli cell of the testes. There is much evidence, both direct and indirect, to support this contention. First of all, ABP is found in seminiferous tubules isolated from rat testis (Tindall *et al.*, 1977). It is evident from ligation experiments, where ABP was found to accumulate in rat testes following ligation of the efferent ducts (French and Ritzen, 1973a), that ABP is a secretory protein, which is transported from the seminiferous tubule to the epididymis via the efferent ducts. Binding activity was not found in lymphatic fluid isolated from the testes, in Leydig cell preparations, or in rat serum in the initial discriptions of ABP origin. Recent evidence suggests that ABP does leak into the circulation (Gunsalas *et al.*, 1978). However, it disappears from the blood after castration. Taken together, these data suggest that ABP is produced within a cell (or cells) in and residing within the testis seminiferous tubule.

Germ cells represent the largest population of cells within the seminiferous tubule. This entire population of cells can be eliminated from rat testes by irradiation of the fetuses *in utero* with a low dosage (125 R) of whole body radiation (Means and Huckins, 1974). When the gonocytes attempt the definitive postnatal division, they regress completely. Thus, the seminiferous tubules of these irradiated rats consist primarily of Sertoli cells surrounded by the peritubular matrix, and the testis are considered to be Sertoli cell-enriched (SCE). When SCE testes were examined for ABP, a distinct peak of androgen binding activity was observed in polyacrylamide gels which coincided with ABP from normal testes (Tindall *et al.*, 1974). The transport of this protein to the epididymis was suggested by similar analyses of cytosol from the epididymis, where higher concentrations were found in the caput than in the cauda. The ABP from the SCE testes was clearly not androgen receptor, since it was stable to treatment with heat (50°C for 30 minutes), sulfhydryl reagents (*N*-ethylmaleimide), and chronic treatment with charcoal (6 hours). The half-time of dissociation of DHT from this protein was 6 minutes, again demonstrating its similarity to ABP and differentiating it from the androgen receptor. Further support of these findings was obtained from irradiating testes of adult rats with high levels (580 R) of irradiation (Hagenas *et al.*, 1975). Following such treatment most tubules are devoid of germ cells after 30 days. In spite of the disappearance of germ cells from these testes, ABP was still present in large amounts. These studies revealed that ABP was produced in seminiferous tubules containing essentially only Sertoli cells.

Androgen binding protein has also been described in testes of rats and mice with genetic defects that result in abnormal spermatogenesis. Testicular feminized (Tfm) rats have bilateral inguinal testes with well developed Leydig cells but arrested spermatogenesis. This animal is characterized by an insensitivity to androgens and lack of differentiation of accessory sex organs due to a paucity of androgen receptors (Ritzen *et al.*, 1971; Bardin *et al.*, 1973; Griffin and Wilson, 1980). ABP levels in the Tfm testes are significantly higher than in the normal testis, whether expressed per milligram protein or per testis (Hansson *et al.*, 1978). However, the secretory rate of ABP *in vitro* is significantly lower in the Tfm compared to the normal. Thus, the increased levels of ABP found in the Tfm testis *in vivo* probably reflect an accumulation of ABP due to atresia of the efferent ducts rather than an elevated production. Nevertheless, the presence of ABP in these testes, where germ cell populations are depleted, is further evidence that ABP is produced by the Sertoli cell.

Another genetically altered testis, which is devoid of germ cells, is that from adult SXR mice. Androgen binding in seminiferous tubules from SXR mice was similar to levels found in normal mice (C3H-101-hcG strain) when expressed per milligram protein but 80% less when expressed per testes (Fritz *et al.*, 1974). It should be pointed out that the only property used to characterize these androgen binding proteins was the K_d determined by a charcoal binding assay. These values were quite different, ranging from 0.6–0.9 nM in the SXR to 1.8-2.7 nM in the normal. Therefore it is difficult to tell if ABP was being measured in both instances or if some other binding component such as receptor protein was being detected.

Normal levels of ABP were reported to be present in cryptorchid testes, where germ cells had been depleted because of heat (Vernon *et al.*, 1974). Initially, this observation was interpreted as indirect evidence that ABP was being produced by Sertoli cells. However, closer examination of production rate and secretion of ABP from cryptorchid testes revealed a severe impairment of Sertoli cell function (Hagenas and Ritzen, 1976). Although no major changes were found in the concentration of ABP per milligram protein, there was a significant decrease in ABP content per testis. Also, the rate of ABP production as measured both *in vivo* and *in vitro* was markedly decreased. Finally, both nitrofurazone and ethionine are known to damage the seminiferous epithelial cells. Following treatment with these drugs, there was an almost complete destruction of germinal elements within the seminferous tubules. However, the rate of ABP production was found to be diminished only slightly compared to controls (deKretzer *et al.*, 1979).

Direct evidence that ABP is produced by the Sertoli cell has been obtained by several laboratories using Sertoli cells maintained in primary culture (Steinberger *et al.*, 1975; Fritz *et al.*, 1976; Tindall *et al.*, 1977; Welsh *et al.*, 1980; Lipshultz *et al.*, 1982; Tindall *et al.*, 1983). There are two published techniques for isolating Sertoli cells. One technique uses trypsin (0.25%) to separate seminiferous tubules from interstitial tissue followed by collagenase (0.1–0.25%) to disengage the Sertoli cells from the tubules (Steinberger *et al.*, 1975; Fritz *et al.*, 1976). The other technique utilizes an initial treatment with collagenase (0.02%), that causes interstitial fragments to aggregate into a mass that can be removed with forceps (Welsh and Wiebe, 1975). This is followed by treatment with pancreatin (0.02%), which causes the peritubular cells to aggregate into one or several masses that again can be removed with forceps. Both of these techniques are satisfactory for isolating Sertoli cell clusters from testes. However, contaminating germ cells remain regardless of the technique used and can be removed only by culturing for a period of time. Since the Sertoli cells attach to the culture dish and the germ cells do not, the germ cells are eventually removed with media changes. Problems with germ cell contamination have been overcome by using Sertoli cell-enriched testes. In this case the enzymatic treatment takes only approximately 2 hours and yields a pure preparation of Sertoli cells (Tindall *et al.*, 1977). The ease, speed, and purity of these cell preparations are of tremendous advantage when acute hormone effects on Sertoli cells are being assessed.

When isolated Sertoli cells from Sertoli cell-enriched rat testes were examined for androgen binding, androgen receptor was found to bind approximately 33 fmol of [^{3}H]testosterone per 10^6 cells under saturating conditions, whereas ABP was much higher (approximately 107 fmol per 10^6 cells) (Tindall et al., 1977). Sertoli cells grown in primary culture will secrete androgen binding protein. However, in the absence of serum or other trophic factors, the amount of ABP recovered decreased with time. Even in the presence of trophic stimuli, a decreased daily secretion of ABP was found. Thus, it is quite evident that androgen binding protein is produced by Sertoli cells, but it is unclear whether synthesis of ABP continues in vitro or whether ABP already synthesized is being secreted. Only by labeling the ABP with radioactive amino acids and precipitating the synthesized produce with a monospecific antibody can this question be answered.

To determine whether an androgen binding protein-like molecule were secreted by the human Sertoli cells, serum free-culture medium was concentrated by molecular filtration, incubated with [^{3}H]DHT and applied to a DEAE cellulose column (Lipshultz et al., 1982; Tindall et al., 1983). An androgen binding component was retained by the DEAE cellulose and this binding activity was eluted with 0.05–0.1 M KCl. This was the same salt concentration required to elute human epididymal ABP (Lipshultz et al., 1977) and rat epididymal ABP (Tindall and Means, 1980) from DEAE cellulose. The binding sites were specific and saturable since a 100-fold excess of unlabeled DHT displaced all of the radioactive DHT. Further characterization was achieved by application of the concentrated medium to a hydroxylapatite column. The androgen binding component from human Sertoli cell was adsorbed by HAP and was eluted with very low concentrations (<30 mM) of potassium phosphate. This is also the phosphate concentration required to elute rat ABP from hydroxylapatite (Tindall and Means, 1980). Again the binding sites were saturable with excess unlabeled DHT. Finally, the migration of the human ABP on steady-state polyacrylamide gels was examined. Androgen binding activity from human Sertoli cells migrated on the gel with an R_f of 0.38, similar to that of human epididymal androgen binding activity (0.35) (Lipshultz et al., 1977), but slightly less than that observed for rat Sertoli cell ABP (0.50) (Tindall and Means, 1980). Excess unlabeled DHT in the gel competed for binding with labeled steroid. These data were the first to demonstrate that human Sertoli cell synthesizes and secrete an androgen binding protein.

The concentration of ABP are highest in the initial segment of the caput epididymis of the rat and decrease with progression from the caput to the corpus to the cauda. No ABP has been detected in the vas deferens of the rat and presumably no ABP is found in rat ejaculates, although this has not been measured. However, in ram seminal plasma, large concentrations of ABP have been demonstrated (Jegou et al., 1978). ABP was also detected in seminal plasma from the billy goat and the bull, whereas, no ABP was detected in ·eminal plasma from human, pig, or horse (Jegou *et al.*, 1978). In the ram the concentration of ABP measured in rete testis fluid is higher in the breeding season than in the nonbreeding season (4.4 × 10^{-9} vs 2.6 × 10^{-9} *M*). The affinity constant did not change (Jegou *et al.*, 1978).

The physicochemical properties of ABP have been characterized extensively. The molecular weight of ABP calculated from the amino acid composition is 105,000 (Tindall and Means, 1980), which is in close agreement with that calculated from gel filtration (Stokes radius: 4.7 nm) and sedimentation analysis (sedimentation coefficient = 4.6 S) of 94,000. Similar values have been reported for rat ABP using gel filtration data (Hansson, 1972) or gel electrophoresis using Ferguson plots (Ritzen *et al.*, 1973). A somewhat smaller molecular weight (73,000) has been reported for rabbit ABP (Hansson *et al.*, 1974). Following denaturation with SDS rat ABP separates into subunit of 47,000 and 41,000 (Musto *et al.*, 1980; Taylor *et al.*, 1980; Larrea *et al.*, 1981). The ratio of the 47,000 to 41,000 subunits is 3:1. Evidence that the 41,000 subunit is part of the native molecule and not a breakdown product was demonstrated using a cross-linking reagent, dimethylsuberimidate. After treatment with this reagent a 94,000-MW protein appeared on an SDS gel with a concomitant decrease of both 47,000 and 41,000 components. Rabbit ABP also appears to be composed of subunits. Subunits of 47,000 and 43,000 MW have been reported (Cheng and Musto, 1982).

The isoelectric point of ABP is 4.6 for the native protein. When the subunit of ABP were separated by SDS polyacrylamide gel electrophoresis and then electrofocused in a second dimension, multiple isoelectric species of both the 47,000 and 41,000 proteins were demonstrated. Whether or not these represent phosphorylated species or carbohydrate residues remains to be determined. There appears to be a loss of the more acidic species and an appearance of the more basic species of ABP as this protein moves from the testis to the cauda epididymis (Larrea *et al.*, 1981).

B. γ-GLUTAMYL TRANSPEPTIDISE

γ-Glutamyl transpeptidase (γ-GTP), a membrane bound enzyme, has a wide spread distribution in human and other mammalian tissues (Sikka and Kabra, 1980). It catalyzes the transfer of the γ-glutamyl moiety of glutathione and other γ-glutamyl compounds to a large number of amino acid and peptide cofactors. It is thought that this enzyme plays a role in the transport of the amino acid via the γ-glutamyl cycle. According to this proposal γ-GTP mediates the translocation of amino acids across the plasma membrane by interacting with both the extracellular amino acid and intracellular glutathione.

γ-GTP may be a testicular marker of Sertoli cell function in the rat (Hodgen and Sherins, 1973; Krueger *et al.*, 1974). The specific activity of γ-GTP increased coincident with cessation of Sertoli cell mitosis and just before the onset of primary spermatocyte formation. Enzyme levels remained constant during the completion of spermatogenesis (Hodgen and Sherins, 1973). In vitamin A-deficient rats, testicular levels of γ-GTP were high, despite virtual complete germ cell depletion, although the activity of lactate dehydrogenase and sorbitol dehydrogenase fell in proportion to the reduced number of advanced germ cells. In man there is a persistance of high testicular levels of γ-GTP activity, despite germ cell depletion, and a low level of this enzyme, when there is profound immaturity of seminiferous tubules and Sertoli cells (Sherins and Hodgen, 1976). In the rat the specific activity of γ-GTP increases with age until day 15 at which time a plateau is reached (Hodgen and Sherins, 1973). This plateau coincides with the cessation of mitotic activity of Sertoli cells.

Within the rat γ-GTP activity appears to be correlated solely with the Sertoli cell. Different cell types within the rat testis were isolated and examined for γ-GTP activity. Isolated Sertoli cells contained very high concentrations of γ-GTP activity whereas germ cells contained less than 10% of the activity associated with the Sertoli cells and may have been due to contamination of Sertoli cells in the germ cell preparation (Lu and Steinberger, 1977). Peritubular cells and interstitial cells contained nondetectable levels of γ-GTP activity. In Sertoli cells isolated from human testes the specific activity of γ-GTP was very similar to the rat Sertoli cell (Lipshultz *et al.*, 1982; Tindall *et al.*, 1983). Interestingly, no γ-GTP activity was found in a cell line derived from a Leydig cell tumor. Thus, γ-GTP appears to be an important biological marker for Sertoli cells within the testis of both humans and rats.

C. TRANSFERRIN

Transferrin is a serum protein found in all vertebrates. It is a single glycopeptide of a reported molecular weight of 70,000–80,000 and functions as an iron transport protein. One of the major proteins secreted by rat Sertoli cells has a molecular weight on SDS gels which corresponds to rat serum transferrin (Skinner and Griswold, 1980). This protein bound to ^{59}Fe and had electrophoretic properties on a native polyacrylamide gel similar to serum transferrin. It was precipitated by a serum transferrin antibody, and antibodies against Sertoli cell secretory proteins coprecipitated with purified rat transferrin. It has been proposed that the Sertoli cell transferrin serves as a source of iron for the heme proteins or for nonheme metaloproteins in developing germ cells. Thus the Sertoli cell could serve as an intermediate in the transport of iron from serum transferrin to the germinal cells in the adluminal compartment. It has been reported by Thorbecke *et al.* (1973), who surveyed several tissues with transferrin antibodies, that transferrin is synthesized by both the testis and the ovary. The significance of the testicular transferrin can only be postulated and further studies of iron metabolism during spermatogenesis are needed.

D. Inhibin

The secretion of follicle stimulating hormone and leuteinizing hormone by the pituitary gland is regulated in part by gonadal substances. While the gonadal steroids are mainly involved in the control of LH, the regulation of FSH secretion appears to involve also a nonsteroidal, water-soluble substance, termed "inhibin." An inhibin-like substance, which selectively suppresses FSH synthesis and release, has been shown to be secreted by rat Sertoli cells (Steinberger and Steinberger, 1976; Chowdhury et al., 1978). In Vitro experiments have demonstrated that a Sertoli cell inhibitor-like factor (SCF) and other inhibin preparations can affect FSH secretion by acting directly on anterior pituitary cells (Steinberger and Steinberger, 1976; Chowdhury et al., 1978; Labrie et al., 1978; DeJong et al., 1979; Franchimont et al., 1979; Scott et al., 1980). Recently, Steinberger et al. (1982) demonstrated the presence of inhibin like SCF binding sites in the rat anterior pituitary. The binding to pituitary homogenates was saturable, time and temperature dependent, and increased with increasing pituitary homogenate concentration. The binding of [^{3}H]SCF was specific since it was inhibited in a dose-related manner by unlabeled SCF but not by several proteinaceous substances. The binding to rat anterior pituitary was also organ specific, as it was 5–6 times greater as compared to equivalent amount of rat brain or spleen and about 3 times greater as compared to liver, kidney, or testis.

IV. Hormonal Regulation by FSH

Follicle stimulating hormone binds to specific receptors on the Sertoli cell plasma membrane, stimulates adenylate cyclase, activates soluble cAMP-dependent protein kinase, and stimulates RNA and protein synthesis (Means and Tindall, 1975). By affecting these intracellular components, FSH regulates specific protein secretion and cell motility.

A. FSH Receptors and Mechanism of Action

High-affinity, low-capacity receptors specific for FSH have been described on Sertoli cell membranes by a number of laboratories (Means and Vaitukaitis, 1972; Bhalla and Reichert, 1974; Desjardins et al., 1974; Rabin, 1974). That these receptors mediate the FSH response is suggested by the dissociation constant near the physiological concentration of FSH; the time course of binding that parallels stimulation of adenylyl cyclase but preceeds other biological responses of the Sertoli cell to FSH; and the observation that the Sertoli cell contains the bulk, if not all of the testicular receptors for FSH. A temporal relationship has been established between binding of FSH to isolated membranes prepared from Sertoli cells (Cheng, 1975; Abou-Issa and Reichert, 1977; Orth and Christiansen, 1977), activation of adenylyl cyclase (Van Sickle *et al.*, 1981), and inhibition of cyclic nucleotide phosphodiesterase (Fakunding *et al.*, 1976). These altered enzyme activities have been shown to result in an elevation in the intracellular levels of cAMP in whole cell preparations from testes of immature rats (Murad *et al.*, 1969; Braun and Sepsensol, 1974; Dorrington and Fritz, 1974), hypophysectomized animals (Kuehl *et al.*, 1970), cryptorchid animals (Dorrington and Fritz, 1974), X-irradiated animals (Means and Huckins, 1974; Fakunding *et al.*, 1976), and isolated Sertoli cells in tissue culture (Dorrington *et al.*, 1975; Heindel *et al.*, 1975; Van Sickle *et al.*, 1981). Cyclic AMP binds to the regulatory subunit of protein kinase, thereby releasing the catalytic subunit of protein kinase. Free catalytic subunit can phosphorylate a number of proteins within the cell. The resulting enhanced protein phosphorylation is followed by increased synthesis of nuclear RNA and stimulated protein synthesis (Means and Tindall, 1975).

Protein kinase is regulated by a small-molecular-weight protein which inhibits cyclic AMP-dependent protein kinase (Beale *et al.*, 1977a,b). Recent evidence has suggested a regulatory role of the protein kinase inhibitor (PKI) in the rat testis Sertoli cell; significant changes in the specific activity of PKI were observed following hypophysectomy and subsequent treatment with FSH (Beale *et al.*, 1977a). The protein has been purified to homogeneity from rat testis (Beale *et al.*, 1977b) and a polyclonal antibody has been raised against this protein (Tash *et al.*, 1980). The cAMP-dependent PKI in Sertoli cell-enriched rat testes is specifically stimulated by FSH *in vivo* (Tash *et al.*, 1979). Stimulation of inhibitor activity was both dose dependent and FSH specific. Near maximal enhancement of PKI activity was achieved with 5 µg purified FSH. Hormonal stimulation of PKI is dependent upon continual protein synthesis; cycloheximide completely prevents the hormone-dependent increases in inhibitor content. PKI is specifically regulated by hormones in culture of rat Sertoli cells maintained under completely defined conditions. Hormones that are known to elevate Sertoli cell cAMP concentrations, namely FSH and isoproterinol, produce a 4- to 5-fold increase in the specific activity of PKI, whereas testosterone and LH have no effect. The stimulating effects of FSH or isoproterinol on PKI are completely mimicked by dibutyryl cAMP. Using specific antibodies to PKI, Tash *et al.* (1981) demonstrated that this protein is regulated by hormones, via preferential stimulation of *de novo* synthesis of the protein. Thus PKI is an important intracellular marker for Sertoli cell function.

Human Sertoli cells are also regulated by FSH *in vitro* (Lipshultz *et al.*, 1982;

Tindall *et al.*, 1983). These cells, isolated and grown in culture, contain γ-glutamyl transpeptidase (γ-GTP) activity similar to that of the rat Sertoli cell. A number of laboratories have demonstrated that rat Sertoli cell γ-GTP is unresponsive to FSH treatment (Lu and Steinberger, 1977; Lipshultz *et al.*, 1982; Tindall *et al.*, 1983). In contrast to the rat Sertoli cell, γ-GTP in the human Sertoli cell does respond to FSH (Lipshultz, 1982; Tindall *et al.*, 1983). This response is hormone specific. Neither LH nor testosterone alone stimulates γ-GTP activity. However, FSH caused a 2-fold increase in γ-GTP activity, and this increase was not altered by the simultaneous presence of testosterone. The FSH stimulation appears to be mediated by cAMP, since dibutyryl cAMP at a concentration of 0.1 μg/ml stimulated γ-GTP activity to the same extent as did FSH. The FSH response was observed only in the continual presence of serum in the culture medium. This FSH-mediated increase in γ-GTP activity was dependent on the amount of hormone. An increase in γ-GTP activity is detected with as little as 0.05 μg FSH/ml. γ-GTP activity increases as a function of FSH concentration, up to 0.5 μg/ml and begins to plateau at about 5 μg/ml of hormone. FSH concentration greater than 5 μg/ml give no further response. Thus, the γ-GTP response appears to occur in the range of FSH concentrations (0.05–0.30 μg/ml) normally found in circulation (Stearns *et al.*, 1974).

B. Cellular Responses

A number of cellular events within the Sertoli cell are altered in response to stimulation by FSH. Both biochemical and morphological responses to FSH have been observed. All responses to FSH appear to be mediated by cAMP. With the use of monospecific antisera to LH and FSH it has been possible to determine the degree of regulation these hormones demonstrate over the fine structural characteristics of the rat Sertoli cell (Dym and Raj, 1977; Chemes *et al.*, 1979). Following treatment with LH antisera the characteristic infoldings of the nuclear envelope disappeared and the nucleus became more rounded. The euchromatic profile of the nucleus was changed to one of more peripheral clumps of heterochromatin. The number of nuclei was reduced as was the number of mitochondria, SER, and Golgi profiles. Moreover, this treatment caused an increase in the number of lipid inclusions, which is also a characteristic feature in aging and degenerating cells of many cell types. The addition of testosterone-propionate to this treatment caused the nucleus to resume its normal characteristics but did not reverse the general atrophied nature of the cytoplasm.

Following treatment with FSH antisera, the Sertoli cell volume was decreased but there appeared to be a normal complement of cytoplasmic organelles. There was not a notable change in nuclear fine structure. This treatment did, however, cause an increase in the dilatation of SER. In both of the treatment regimes, there were no changes in the junctional complexes.

It is also of interest that there are fine structural changes in the Sertoli cell of the seasonally breeding rodent *Rattus fuspcipes* during the winter months when the serum levels of LH, FSH, and testosterone are significantly lower (Hansson *et al.*, 1975). The fine structure alterations are very similar to those demonstrated in the antisera studies.

The fine structure of the Sertoli cell then appears to be more highly regulated by LH than FSH. Much of this regulation is presumably via the stimulatory effect of LH on testosterone production by the testis and the subsequent stimulation of the Sertoli cell by testosterone. However, the addition of testosterone to LH-depleted animals did not completely restore the fine structural details, suggesting there may also be a direct effect on LH on Sertoli cell morphology.

V. Hormonal Regulation by Testosterone

The Sertoli cell also is regulated by the male sex hormone, testosterone. Testosterone stimulates ABP activity *in vivo*. This has been demonstrated both acutely (Tindall and Means, 1976) and chronically (Elkington *et al.*, 1977). However, the acute stimulation appears to result from a stabilization of the protein in the presence of ligand (Tindall *et al.*, 1978), whereas chronic treatment is more closely correlated with production of ABP by the Sertoli cell. Chronic treatment of hypophysectomized rats with testosterone has demonstrated that this steroid will maintain normal levels of ABP in the apparent absence of FSH (Weddington *et al.*, 1976).

The synthesis of ABP is significantly increased over control values when Sertoli cell cultures are incubated with testosterone (Louis and Fritz, 1979; Karl and Griswold, 1980). Evidence from cultured Sertoli cells is compatible with a requirement for androgen receptors for the ABO response (Louis and Fritz, 1979). These studies revealed that the concentration of testosterone measured in the media at half-maximum ABP stimulation was near the K_d of receptor binding, Similarly, cyproterone acetate, which competitively inhibits binding of testosterone to the androgen receptor, prevented the ABP stimulation. Thus these data suggest that a receptor-mediated event may be involved in the production and secretion of ABP by the Sertoli cell. Even though testosterone stimulates the production of ABP both *in vivo* and *in vitro,* this is obviously not the only hormone required, since ABP can be stimulated *in vitro* in the absence of testosterone by other factors. Moreover, the androgen receptor is not essential for the *in vivo* response, since ABP is present in tfm rat testes, which have reduced numbers of androgen receptors. Nevertheless, androgens are essential for the maintenance of spermatogenesis. Androgen appears to act via the Sertoli cell receptor and not via germ cell receptors, since x/y ↔ x t + m/y chimeric mice, which lack androgen receptors in their germ cells, produce normal sperm which are fertile. Thus the androgen receptor in the Sertoli cell plays a pivotal role in the hormonal regulation of spermatogenesis.

High-affinity, low-capacity receptors that are specific for testosterone have been demonstrated in the Sertoli cell. Both cytoplasmic (Tindall *et al.*, 1977) and nuclear (Sanborn *et al.*, 1977) androgen receptors have been described. Additionally, chromatin acceptor sites for androgen receptors have been documented in cultures of rat Sertoli cells (Tsai, 1980; Tsai and Steinberger, 1982). The cytoplasmic receptor exhibits a dissociation constant, K_d, or 7.6×10^{-10} M for testosterone. The receptor can be precipitated quantitatively with 40% ammonium sulfate, by separating it from ABP. The receptor is labile to heat, long-term (18 hour) charcoal treatment, and sulfhydryl reagents, indicating a sulfhydryl containing amino acid close to or within the active binding site. Its dissociation rate is greater than 24 hours and a lower relative mobility than ABP on polyacrylamide gels. The concentration of the receptor in cultured Sertoli cells is 33 fmol/10^6 cells. The Sertoli cell nuclear receptor has a K_d of 2.5×10^{-9} M. Approximately 30% of the nuclear bound hormone is extracted within 1 hour by 0.4 M KCl and 34% of this activity is bound. The nuclear accumulation of androgen is both time and temperature dependent.

In order to define the mechanism by which androgen receptors mediate the biological effect of testosterone within the Sertoli cell, a clear understanding of both the physicochemical and biological properties of this protein is necessary. Preparations of purified androgen receptor have only recently been available (Chang *et al.*, 1982, 1983). There are many difficulties associated with purifying the androgen receptor from the Sertoli cell because of a number of factors, including high concentrations of endogenous androgen, a paucity of Sertoli cells within the testis (~2%), and the low concentration of androgen receptors within the Sertoli cell. However, other accessory sex organs are available which satisfy all of the criteria needed for obtaining large amount of androgen receptor. Both steer seminal vesicle and rat ventral prostate have been utilized for starting source of androgen receptor for purification purposes.

The procedure for receptor purification was developed from a consideration of the biological and physicochemical properties of the receptor molecule. The receptor was purified by a combination of differential DNA chromatography and testosterone affinity chromatography. Differential DNA chromatography is achieved by allowing the nontransformed receptor to pass through a DNA-Sepharose column. The nontransformed receptor does not bind to the column and passes through. This step removes those proteins other than the receptor, which have an affinity for DNA. After this step the receptor is transformed to a DNA-binding state either by heat, high salt, or ammonium sulfate precipitation. The transformed receptor is then applied to a second DNA column, and the receptor binds to the DNA. Other proteins which failed to bind to DNA on the first column also pass through the second column, and are eliminated. The final property of the receptor utilized in the purification was its high affinity for pyridoxal phosphate. Elution of the receptor from the second DNA column with 10 mM pyridoxal phosphate resulted in an overall purification of greater than 100,000-fold. Approximately 3–8 μg of receptor protein was obtained from 35 g of tissue, with a yield of 24–48%. The purified receptor from rat ventral prostate had a molecular weight of 86,000; whereas 60,000–70,000 MW species were obtained from steer seminal vesicle. The differences in molecular weight from these two tissues could be due to several possibilities, proteolysis being one. The tissue from the slaughter house took longer to process than that obtained in the laboratory. Moreover, the steers had been castrated for up to 1 year, which markedly altered the tissue morphology. Indeed, histology of the tissue revealed a reduced epithelial cell population, and mostly stromal elements were present. Finally, an obvious possible reason is species differences. Nevertheless, it is interesting that these receptors did have different molecular weights.

Two affinity labels were used to confirm the molecular weights of the purified receptors. Dihydrotestosterone 17β-bromoacetate is an alkylating affinity label which binds covalently to the androgen receptor. [^{3}H]DHT bromoacetate was incubated with a preparation of partially purified receptor. Fluorography of the SDS gels showed one band of radioactivity at 86,000 daltons for the prostate receptor and 60,000–70,000 daltons for the seminal vesicle receptor, thus confirming the molecular weights of the purified proteins. Radioactive DHT bromoacetate was displaced with excess unlabeled ligand, whereas no competition was observed with progesterone, 17β-estradiol, dehydroepiandrosterone, or 5α-androstane-3α-17β-diol, indicating that the binding component were saturable and steroid specific. Similar results were obtained using a photoaffinity label, ^{3}H-R1881.

One feature common to all androgen receptors studied to date is their charge properties. Chromatofocusing of either crude or purified preparations has demonstrated similar isoelectric points of 6.3–6.8 for androgen receptors in steer seminal vesicle, rat prostate, rat seminal vesicle, Dunning prostatic tumor, and human BPH. The pI of the androgen receptor is different from the ABP (pI = 4.6) (Tindall et al., 1977) and SBP (pI = 4.6) (Tindall and Means, 1980).

The most unique property of the androgen receptor in the Sertoli cell is its specificity of binding to the steroid. The steroid specificity of the Sertoli cell receptor is similar to that in the rat ventral prostate and human foreskin fibroblast (Tindall et al., 1977; Cunningham et al., 1981, 1983). The C-3 ketone and the C-17 hydroxy group appear to be critical for this binding specificity. There are several alterations in the steriod molecule, which will increase the affinity of the steroid for the receptor. Removal of the C-19 methyl group causes a marked increase in affinity of both testosterone and dihydrotestosterone for the androgen receptor. The increased affinity probably results from a decreased stearic hindrance imposed by the methyl group and a resulting planarity of the steroid, thus a better fit in the active binding site. Interestingly the addition of methyl groups at both the C-7 and the C-17 position results in increased affinity for the receptor. Why this is so is not known, but may involve hydrogen bonding between the methyl groups and the receptor site. Further studies using modified steroids should provide insights for chemists to synthesize more effective androgens, and possibly antiandrogens.

VI. Conclusions

We can conclude from the above evidence that the Sertoli cell plays an important role in spermatogenesis. Clearly, the Sertoli cell can mediate the biological actions of both FSH and testosterone. It is logical to conclude that cellular products synthesized in the Sertoli cell under the control of FSH and testosterone can act on germ cells to influence their development. A number of proteins which are synthesized by the Sertoli cell have been identified during the past few years. The role(s) these proteins play in spermatogenesis, if any, remains to be determined. Nevertheless, scientists are currently in a unique position of analyzing the secretory products of the Sertoli cell in hopes of obtaining those products which are essential to the spermatogenic process. Recently germ cells have been cocultured with both Sertoli and peritubular cells. Also, segments of seminiferous tubules have been cultured *in vitro* for up to 8 days and completion of meiotic divisions has been described. These isolated systems should aid us in disecting out the many complicating factors involved in spermatogenesis and the role Sertoli cells play in this process.

REFERENCES

Abou-Issa, H., and Reichert, L. E., Jr. (1977). *J. Biol. Chem.* **252,** 4166–4174.
Bardin, C. W., Bullock, L. P., Sherins, R. J., Mowscowicz, I., and Blackburn, W. R. (1973). *Recent Prog. Horm. Res.* **29,** 65–109.
Bawa, S. R. (1963). *J. Ultrastruct. Res.* **9,** 459–474.
Beale, E. G., Dedman, J. R., and Means, A. R. (1977a). *J. Biol. Chem.* **252,** 6322–6327.
Beale, E. G., Dedman, J. R., and Means, A. R. (1977b). *Endocrinology* **101,** 1621–1634.
Bhalla, V. K., and Reichert, L. E., Jr. (1974). *J. Biol. Chem.* **249,** 43–51.
Black, V. H. (1971). *Am. J. Anat.* **131,** 415–426.
Braun, T., and Sepsensol, S. (1974). *Endocrinology* **94,** 1028–1033.
Chang, C. H., Rowley, D. R., Lobl, T. J., and Tindall, D. J. (1982). *Biochemistry* **21,** 4102–4109.
Chang, C. H., Rowley, D. R., and Tindall, D. J. (1983). *Biochemistry* **22,** 6170–6175.
Chemes, H. E., Dym, M., and Raj. H. G. M. (1979). *Biol. Reprod.* **21,** 251–262.
Cheng, Kwong-Wah (1975). *Biochem. J.* **149,** 123–132.
Cheng, S. L., and Musto, N. A. (1982). *Biochemistry* **21,** 2400–2405.

Chowdhury, M., Steinberger, A., and Steinberger, E. (1978). *Endocrinology* **103**, 644–647.
Christensen, A. K. (1965). *Anat. Rec.* **151**, 335.
Christensen, A. K., and Fawcett, D. W. (1966). *Am J. Anat.* **118**, 551–572.
Clermont, G., McCoshen, J., and Hermo, L. (1980). *Anat. Rec.* **196**, 83–99.
Coleman, O. D., and Stockhert, J. C. (1978). *Microsc. Acta* **81**, 27–30.
Connell, C. J. (1978). *J. Cell Biol.* **76**, 57–75.
Cunningham, G. R., Tindall, D. J., Lobl, T. J., Campbell, J. A., and Means, A. R. (1981). *Steroids* **38**, 243–261.
Cunningham, G. R., Lobl, T. J., Cockrell, C., Shao, T. C., and Tindall, D. J. (1983). *Steroids* **41**, 617–626.
DeJong, F. H., Smith, S. D., and Vander Molen, H. J. (1979). *J. Endocrinol.* **80**, 91–102.
deKretzer, D. M., Kerr, J. B., Rich, K. A., Risbridger, G., and Dobos, M. (1979). *In* "Comparative Aspects of Testicular Development Structure and Function" (A. Steinberger and E. Steinberger, eds.), pp. 107–116. Raven, New York.
Desjardins, C., Zeleznik, A. J., Midgley, A. R., and Reichert, L. E., Jr. (1974). *In* "Hormone Binding and Target Cell Activation in Testis" (M. L. Dufau and A. R. Means, eds.), pp. 221–236. Plenum, New York.
Dorrington, J. H., and Fritz, I. B. (1974). *Endocrinology* **94**, 395–403.
Dorrington, J. H., Roller, N. F., and Fritz, I. B. (1975). *Mol. Cell. Endocrinol.* **3**, 57–70.
Dorrington, J. H., Fritz, I. B., and Armstrong, D. T. (1978). *Biol. Reprod.* **18**, 55–64.
Dym, M., and Fawcett, D. W. (1970). *Biol. Reprod.* **3**, 308–326.
Dym, M., and Madhwa Raj, H. G. (1977). *Biol. Reprod.* **17**, 676–696.
Elkington, J. S. H., Sanborn, B. M., Martin, M. W., Chowdhury, A. K., and Steinburger, E. (1977). *Mol. Cell. Endocrinol.* **6**, 203–209.
Fakunding, J. L., Tindall, D. J., Dedman, J. R., Mena, C. R., and Means, A. R. (1976). *Endocrinology* **98**, 392–402.
Fawcett, D. W. (1956). *Anat. Rec.* **124**, 401.
Fawcett, D. W. (1975). *Handb. Physiol. Sect. 7* **V**, 21–56.
Fawcett, D. W., and Burgos (1956). *Anat. Rec.* **124**, 401–402.
Fawcett, D. W., Leak, L. V., and Heidger, P. M. (1970). *J. Reprod. Fertil. (Suppl.)* **10**, 105–122.
Flickinger, C. J. (1967). *Z. Mikrosk. Anat. Forsch.* **78**, 92.
Franchimont, P. J., Verstraelen-Proyard, J., Hazee-Hagelstein, M. T., Renard, C. H., Demoulin, A., Bourguignon, J. P., and Hustin, J. (1979). *Vit. Horm.* **37**, 243–302.
Franke, W. W., Grund, C., and Schmid, E. (1979). *Eur. J. Cell Biol.* **19**, 269.
French, F. S., and Ritzen, E. M. (1973). *J. Reprod. Fertil.* **32**, 479–483.
Fritz, I. B., Kopec, B., Lam, K., and Vernon, R. G. (1974). *In* "Hormone Binding and Target Cell Activation in the Testis" (M. L. Dufau and A. R. Means, eds.), pp. 311–327. Plenum, New York.
Fritz, I. B., Rommertz, F. G., Louis, B. G., and Dorrington, J. H. (1976). *J. Reprod. Fertil.* **46**, 17–24.
Gravis, C. J. (1979). *Z. Mikrosk. Anat. Forsch.* **93**, 321–342.
Griffin, J. E., and Wilson, J. D. (1980). *N. Engl. J. Med.* **302**, 198–209.
Gunsalas, G. L., Musto, N. A., and Bardin C. W. (1978). *Science* **200**, 65–66.
Hagenas, L., and Ritzen, E. M. (1976). *Mol. Cell. Endocrinol.* **4**, 25–34.
Hagenas, L., Ritzen, E. M., Ploen, L., Hansson, V., French, F. S., and Nayfeh, S. N. (1975). *Mol. Cell. Endocrinol.* **2**, 339–350.
Hagenas, L., Ploen, L., and Ekwall, H. (1978). *J. Endocrinol.* **76**, 87–91.
Hansson, V. (1972). *Steroids* **20**, 575–594.
Hansson, V., Ritzen, E. M., Weddington, S. C., McLean, W. S., Tindall, D. J., Nayfeh, S. N., and French, F. S. (1974). *Endocrinology* **95**, 690–700.

Hansson, V., Weddington, S. C., and Petrusz, P. (1975). *Endocrinology* **97**, 469–473.
Hansson, V., Purvis, K., Attramadal, A., Torjesen, P., Anderson, D., and Ritzen, E. M. (1978). *Int. J. Androl.* **1**, 96–104.
Heindel, J. J., Rothernberg, R., Robison, G. A., and Steinberger, A. (1975). *J. Cyclic Nucleotide Res.* **1**, 69–79.
Hodgen, G. D., and Sherins, R. J. (1973). *Endocrinology* **93**, 985–989.
Jegou, B., and LeGac-Jegou, F. (1978). *J. Endocrinol.* **77**, 267–268.
Jegou, B., Dacheux, J. L., Tergui, M., Garnier, D. H., and Courot, M. (1978). *Mol. Cell. Endocrinol.* **9**, 335–346.
Karl, A. F., and Griswold, M. D. (1980). *Biochem. J.* **186**, 1001–1003.
Khan, M. S., Knowles, B. B., Aden, D. P., and Rosner, W. (1981). *J. Clin. Endocrinol. Metab.* **53**, 448–449.
Kierszenbaum, A. L., Feldman, M., Lea, O., Spruill, W. A., Tres, L. L., Petrusz, P., and French, F. S. (1980). *Proc. Natl. Acad. Sci. U.S.A.* **77**, 5322–5326.
Krimer, D. B. (1977). *Arch. Biol. (Brussels)* **88**, 117.
Krueger, P. M., Hodgen, G. D., and Sherins, R. J. (1974). *Endocrinology* **95**, 955–962.
Keuhl, F. A., Patanelli, D. J., Tarnoff, J., and Humes, J. L. (1970). *Biol. Reprod.* **2**, 154–163.
Labrie, F., Legace, L., Ferland, L., Kelley, P. A., Drouin, J., Masicotte, J., Bonne, C., Raynaud, J. P., and Dorrington, J. (1978). *Int. J. Androl. Suppl.* **2**, 81–101.
Larrea, F., Musto, N. A., Gunsalus, G. L., and Bardin, C. W. (1981). *Endocrinology* **109**, 1212–1220.
Lipshultz, L. I., Tsai, Y. H., Sanborn, B. M., and Steinberger, E. (1977). *Fertil. Steril.* **28**, 947–951.
Lipshultz, L. I., Murthy, L., and Tindall, D. J. (1982). *J. Clin. Endocrinol. Metab.* **55**, 228–237.
Louis, B. G., and Fritz, I. B. (1979). *Endocrinology* **104**, 454–461.
Lu, C., and Steinberger, A. (1977). *Biol. Reprod.* **17**, 84–88.
Lubarsch, O. (1896). *Anat. Physiol.* **145**, 316.
McGinley, D. M., Posalaky, Z., Porvaznik, M., and Russell, L. (1979). *Tissue Cell* **11**, 741–754.
Means, A. R., and Huckins, C. (1974). *In* "Hormone Binding and Target Cell Activation in the Testis" (M. L. Dufau and A. R. Means, eds.), pp. 145–166. Plenum, New York.
Means, A. R., and Tindall, D. J. (1975). *In* "Hormonal Regulation of Spermatogenesis" (F. S. French, V. Hansson, E. M. Ritzen, and S. N. Nayfeh, eds.), pp. 383–398. Plenum, New York.
Means, A. R., and Vaitukaitis, J. (1972). *Endocrinology* **90**, 39–46.
Murad, F., Strauch, B. S., and Vaughan, M. (1969). *Biochim. Biophys. Acta* **177**, 591–598.
Musto, N. A., Gunsalus, G. L., and Bardin, C. W. (1980). *Biochemistry* **19**, 2853–2860.
Nagano, T. (1966). *Z. Mikrosk. Anat. Forsch.* **73**, 89.
Nagano, T., and Suzuki, F. (1978). *Cell Tissue Res.* **189**, 389–401.
Nicander, L. (1963). *J. Ultrastruct. Res.* **8**, 190–191.
Norris, J. S., and Kohler, P. (1976). *Science* **192**, 898–900.
Orth, J., and Christensen, A. K. (1977). *Endocrinology* **101**, 262–278.
Rabin, D. (1974). *In* "Hormone Binding and Target Cell Activation in Testis" (M. L. Dufau and A. R. Means, eds.), pp. 193–200. Plenum, New York.
Rambourg, A., Clermont, Y., and Hermo, L. (1979). *Am. J. Anat.* **154**, 455–476.
Ritzen, E. M., Nayfeh, S. N., French, F. S., and Aronin, P. A. (1972). *Endocrinology* **91**, 116–124.
Ritzen, E. M., Dobbins, M. C., Tindall, D. J., French, F. S., and Nayfeh, S. N. (1973). *Steroids* **21**, 593–607.
Rowley, D. R., Chang, C. H., and Tindall, D. J. (1983). *Annu. Meet. Endocrine Soc., 65th* Abstr. No. 225.
Russell, L. D. (1979). *Anat. Rec.* **194**, 213–232.

nomenon of loss of peripheral proteins and N-band staining with aging of slides, it is speculated that silver stains the proteins more closely associated with DNA while Giemsa stains peripheral proteins (Taylor and Deleon, 1980). Further, cytochemical studies have shown that Ag-NORs are due to acidic proteins (Schwarzacher *et al.*, 1978; Olert *et al.*, 1979; Kling *et al.*, 1980), and rich in

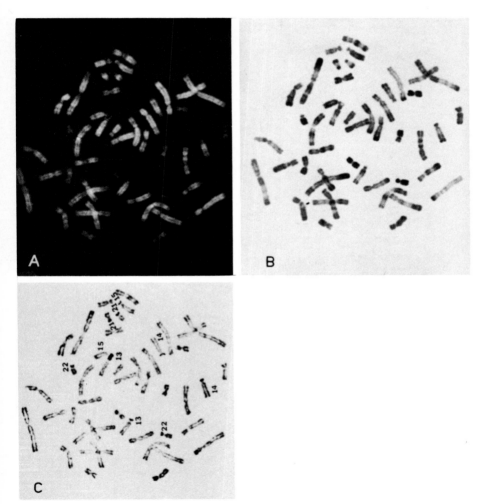

FIG. 1.   Sequentially banded metaphase from a normal female. (A) QFQ (Q bands by flurescence using quinacrine) bands, (B) RFA (R bands by fluorescence using acridine orange) bands, and (C) Ag-NOR bands. RFA chromosomes were photographed in color but printed in black and white (Verma and Lubs, 1975). The identification of acrocentric chromosomes can easily be facilitated by such an approach (Verma and Babu, 1984).

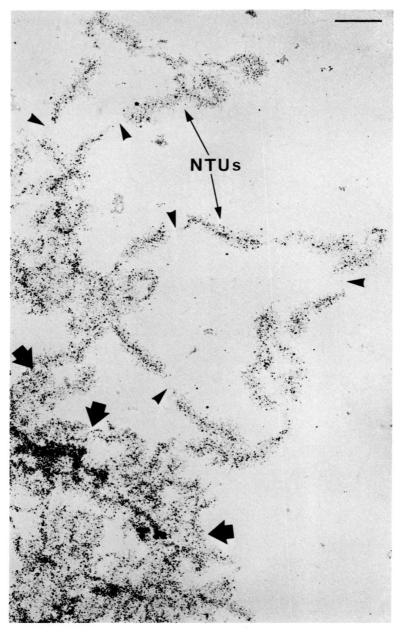

FIG. 2.   General view of the nucleolar transcription units (NTUs) after Ag-NOR staining. Silver grains are located on fibrillar sequences with preferential axial localization. Spacers are devoid of grains (arrows). (Courtesy of Dr. N. Angelier.)

sulfhydryl and disulfide groups (Buys and Osinga, 1980). However, the exact nature of protein is not clear, though they are reported as a single protein (Hubbell et al., 1979), the protein C_{23} and B_{23} (Lischwe et al., 1979; Busch et al., 1982) or large subunit of the RNA polymerase (Williams et al., 1982). The electron miroscopic studies have further revealed that Ag staining is adjacent to NORs and not on NORs (Schwarzacher et al., 1978) and are in contact with the chromatin fibers (Hernandez-Verdun et al., 1982). More recently, Angelier et al. (1982) have shown using molecular spreads of pleurodele oocytes that Ag proteins are present on the nucleolar transcription units (NTUs) but not on spacers (Fig. 2).

The N-band specificity for NORs in some of the plant (Gerlach, 1977) and animal (Faust and Vogel, 1974; Pimpinelli et al., 1976) chromosomes has been questioned while supporting evidence in favor of the method was provided by Funaki et al. (1975). Nevertheless, the applicability of this technique to plant chromosomes still seems to be a matter of controversy (Stack, 1974; Jewell, 1979, 1981).

Some variations of the original silver staining were reported to obtain uniformity (Kodama et al., 1980), to simplify and reduce the timing (Howell and Black, 1980). The modified method of Howell and Black (1980); Verma et al. (1984) is now regularly used in authors laboratory with excellent results (Fig. 1C).

The silver staining alone does not facilitate the precise identification of various chromosomes. Therefore, NOR-silver staining could be supplemented with other stains to visualize the NORs and chromosome bands simultaneously (Lau et al., 1978; Howell and Black, 1978) or sequentially (Bloom and Goodpasture, 1976; Tantravahi et al., 1977; Zankl and Bernhard, 1977; Vedbrat et al., 1979, 1980) (Fig. 1A–C).

III. Distribution of NORs in Metaphase Chromosomes

The ribosomal DNA of different eukaryotes shows close molecular homology but is different from prokaryotes (Sinclair and Brown, 1971). Out of the major rDNA cistrons (18 S and 28 S) is usually clustered at a definite number of sites known as NORs (secondary constrictions) while genes coding for 5 S RNA are distributed over a number of sites in genome (Wimber and Steffensen, 1973; Pardue et al., 1973; Pardue and Birnstiel, 1973). The number and position of NORs vary from species to species and even the classification of karyotypes depending on the NOR distribution pattern was suggested by Hsu et al. (1975). For example, the male species of Corollia perspicillata have single NOR on X chromosome while NORs are distributed over many autosomes in Microtus agrestis (Goodpasture and Bloom, 1975; Hsu et al., 1975).

Though the secondary constrictions are the sites of NORs in general, the other not uncommon sites are "terminal NORs" as in Cricetulus griseus and Corollia *castanea* (Hsu *et al.*, 1975; Goodpasture and Bloom, 1975) and "centromeric NORs" as in *Mus musculus* (Goodpasture and Bloom, 1975), and *Microtus agrestis* (Hsu *et al.*, 1975; Goodpasture and Bloom, 1975). Among the last type of "centromeric NORs" the specific location of NOR in relation to the centromere cannot be demonstrated using conventional preparations. However, using the suitable material, i.e., mouse cell lines which present an unique phenomenon of loss of condensation of centromeric heterochromatin with Hoechst 33258 (Hilwig and Gropp, 1973), Nielsen *et al.* (1979) have further delineated the sites as neighboring, interstitial, and double NORs in relation to centromere.

Some discrepencies are noted in the number of NORs and NOR bearing chromosomes reported by Goodpasture and Bloom (1975) and Funaki *et al.* (1975) which are perhaps due to the different materials and/or methods used in their studies.

IV. NORs in Human Chromosomes

A. MORPHOLOGY

Prior to the cytological demonstration of NORs with specialized techniques the secondary constrictions as less stained regions on short arms of acrocentric chromosomes in human complement were noted (Rothfels and Siminovitch, 1958) and they are usually associated with nucleoli (Ferguson-Smith and Handmaker, 1961; Ohno *et al.*, 1961).

Different number of 18 S and 28 S ribosomal gene copies have been reported per haploid genome, for example, 300 by Jeanteur and Attardi (1969), 220 by Bross and Krone (1972), and 50 by Young *et al.* (1976). These major RNA genes are distributed among specific NOR sites while more number of 5 S genes are distributed over many chromosomes (Hatlen and Attardi, 1971). The biochemical hybridization studies with isolated and separated chromosomes from HeLa cells have suggested that the major rRNA genes are on smaller chromosomes especially D- and G-groups (Huberman and Attardi, 1967). The first cytological localization of rDNA cistrons on short arms of all the acrocentric chromosomes was made by Henderson *et al.* (1972) using the *in situ* hybridization (Pardue and Gall, 1969). Further, the sites of NORs are resolved to be the unstained "stalk" (suggested in Chicago Conference, 1966) but not the distally situated dark regions called satellites (Evans *et al.*, 1974; Goodpasture *et al.*, 1976; Ferraro *et al.*, 1977) (Fig. 3).

B. NATURE OF HUMAN NORs

The number of Ag-positive NORs varies in different individuals from 4 to 10 (Bloom and Goodpasture, 1976; Goodpasture *et al.*, 1976; Varley, 1977;

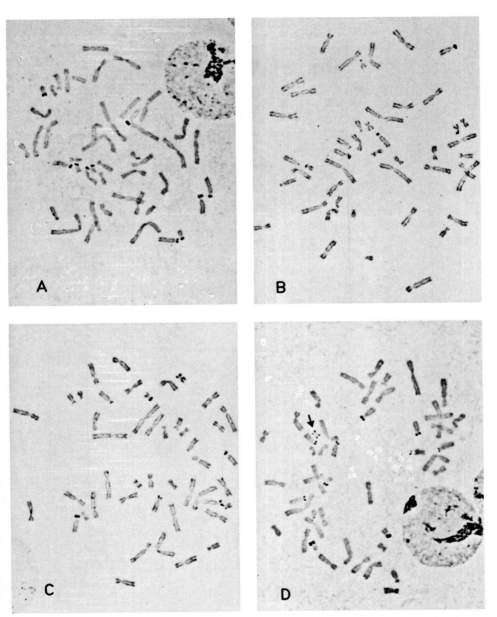

Fig. 3. Metaphase chromosomes from four different individuals showing the variation in number and NORs expressions as demonstrated by Ag-NOR staining. (A) All the 10 acrocentrics show significant silver-positive NORs while (B) and (C) do not have a silver-positive region on all the acrocentrics. (D) Metaphase of an individual with normal phenotype showing an extra bisatellited (arrow) marker chromosome contributing two additional NORs which are also Ag positive and, therefore, presents a total of 12 active NORs.

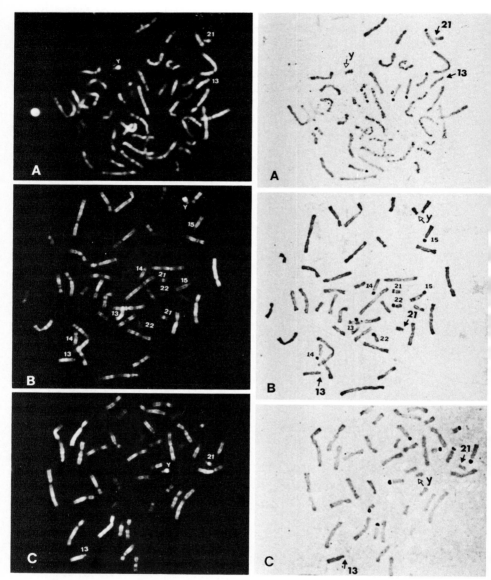

FIG. 4. Chromosomes from three different cells (A to C) of the same individual with various degrees of condensation showing Ag-NOR staining. Each chromosome is preidentified (one of 13 and one of 21) and does not show silver reaction (arrows) and remains the same in all the cells regardless of different condensation stages confirming their inactiveness at different cell cycle stages.

Mikelsaar *et al.*, 1977a) and the amount of Ag reaction varies on each acrocentric chromosome with active NORs of the same individual (Dittes *et al.*,
1975; Warburton *et al.*, 1976; Hayata *et al.*, 1977; D. A. Miller *et al.*, 1977)
(Fig. 4) is proportionate to the amount of rDNA present (Warburton and Henderson, 1979), and consistant from cell to cell of an individual (Howell *et al.*, 1975;
Tantravahi *et al.*, 1977). These characteristics and behavior of these NORs are
heritable (Mikelsaar *et al.*, 1977b; Varley, 1977; Markovic *et al.*, 1978; Vedbrat
et al., 1979; Zakharov *et al.*, 1980, 1982; Egolina *et al.*, 1981; Verma *et al.*,
1982b). The lack of silver staining with all rDNA sites is associated with fewer
cistrons of rDNA, and repressed rDNA (Warburton and Henderson, 1979) or in
other words only the transcriptionally active sites during the preceding interphase
react with silver (D. A. Miller *et al.*, 1976; O. J. Miller *et al.*, 1976). The
evidence for such phenomenal relation was the simultaneous absence of human
rRNA (Eliceiri and Green, 1969) and Ag staining on human acrocentrics in
mouse human somatic cell hybrids (D. A. Miller *et al.*, 1976).

The nucleolar and NOR activity, as demonstrated with silver staining, varies
in different tissues (Arrighi *et al.*, 1980; Reeves *et al.*, 1982). These variations
are attributed to differentiation of cells and their functional requirements. Aug-
mented NOR activity and the increase in the number of nucleoli are noted
following the stimulation of resting cells with phytohemagglutinin and cell fusion (Arrighi *et al.*, 1980).

The metabolic and morphological changes of nucleoli during the malignant
transformation (Busch and Smetana, 1970; Robbins and Angell, 1976) have led
to a number of studies on possible parallel changes in NORs. The clear evidence
of altered NOR activity is reported in leukemic cells (Varley, 1977; Reeves *et
al.*, 1982), hypodiploid cells in meningioma (Zankl *et al.*, 1982), and in ade-
nocarcinoma (Trent *et al.*, 1981). The increase of NOR activity in hypodiploid
and increased number of active NORs in hyperdiploid cells were reported by
Zankl *et al.* (1982) and Trent *et al.* (1981), respectively. Similar increase of
NOR activity is reported in stimulated leukocytes from females with adenocar-
cinoma (Cheng *et al.*, 1981). However, no such increase was observed inspite of
the numerical variations in the number of total and acrocentric chromosomes in
neoplastic cells (Hubbell and Hsu, 1977). Therefore, no particular generalized
pattern could be related with transformation process but change seems to be
specific to individual type and source of tumor.

The NORs in human chromosomes are affected by aging with loss of
hybridizable rDNA (Strehler *et al.*, 1979; Strehler and Chang, 1979) and decreased silver-positive NORs (Buys *et al.*, 1979) which is more pronounced in
females, D-group in specific (Denton *et al.*, 1981). The possibility of decrease in
Ag staining due to age-dependent irreversible repression of rDNA was suggested
by Denton *et al.* (1981). NOR activity also appears to be regulated by hormone

system especially that of thyroid (Hansson, 1971; Nilsson *et al.*, 1975; Zankl *et al.*, 1980).

C. ACROCENTRIC ASSOCIATIONS

The so called satellite associations in human chromosomes described earlier by Ferguson-Smith and Handmaker (1961) have been extensively studied to elucidate various characteristics and factors influencing them. Inspite of the current knowledge that the associations are due to NORs (stalks) but not the satellites, one wonders at the wide use of the term satellite association in the literature. More appropriate terminology like acrocentric association or NOR association is suggested by Verma *et al.* (1983). These associations are due to their participation in nucleolar organization (Ferguson-Smith, 1964) and physical connectives do present between associated chromosomes (Zang and Back, 1969) with rDNA (Henderson *et al.*, 1973).

The association frequency is correlated with the length of stalk region (Evans *et al.*, 1974; Schmid *et al.*, 1974; Zankl and Zang, 1974; Hayata *et al.*, 1977), amount of silver staining (D. A. Miller *et al.*, 1977; Capoa *et al.*, 1978), and the amount of rDNA present in that particular site (Henderson and Atwood, 1976). However, clear correlation between rDNA amount and association frequency is not observed by Evans *et al.* (1974).

The association frequency of acrocentric chromosomes is examined using different criteria (Jacobs *et al.*, 1976; Ardito *et al.*, 1976; Cohen and Shaw, 1967; Zang and Back, 1969; Nankin, 1970; Galperin-Lemaitre *et al.*, 1980; Verma *et al.*, 1983). There are conflicting views on the randomness with which particular chromosomes are involved in NOR associations (Zang and Back, 1969; Patil and Lubs, 1971; Jacobs *et al.*, 1976; Houghton and Houghton, 1978; Hansson, 1979). The number and distribution of associations differ among different individuals (Schmid *et al.*, 1974; Mattei *et al.*, 1976). The chromosome 21 is reported to participate more frequently than others (Ray and Pearson, 1979). The associations contribute to nonrandom arrangement of metaphase chromosomes (Bobrow and Heritage, 1980). Galperin-Lemaitre *et al.* (1980) have even reported the sex-dependent preferential association, i.e., chromosome 21 and 13 participate more frequently than other chromosomes in females.

The acrocentric associations are further classified into "chromatid" and "chromosome" type depending on the involvement of one or both chromatids, respectively (Verma *et al.*, 1982). We have suggested that chromatid type associations are more frequent than chromosome type (Verma *et al.*, 1982). More recently, a relatively new approach with "Giemsa-labeling technique" using 5-bromodeoxyuridine (BUdR) has been employed to investigate the orientation as well as segregation of mitotic chromatids during acrocentric associations (Bobrow and Heritage, 1980). With the help of sister chromatid differentiation techniques some investigators reported random (Strobel *et al.*, 1981; Beek, 1981; Woodruff and Deleon, 1982) while others reported nonrandom segregation of chromatids and association in somatic cells (Bobrow and Heritage, 1980). We have critically observed the lateral orientation of acrocentric chromosomes and found random segregation for single chromatid and nonrandom for double chro-

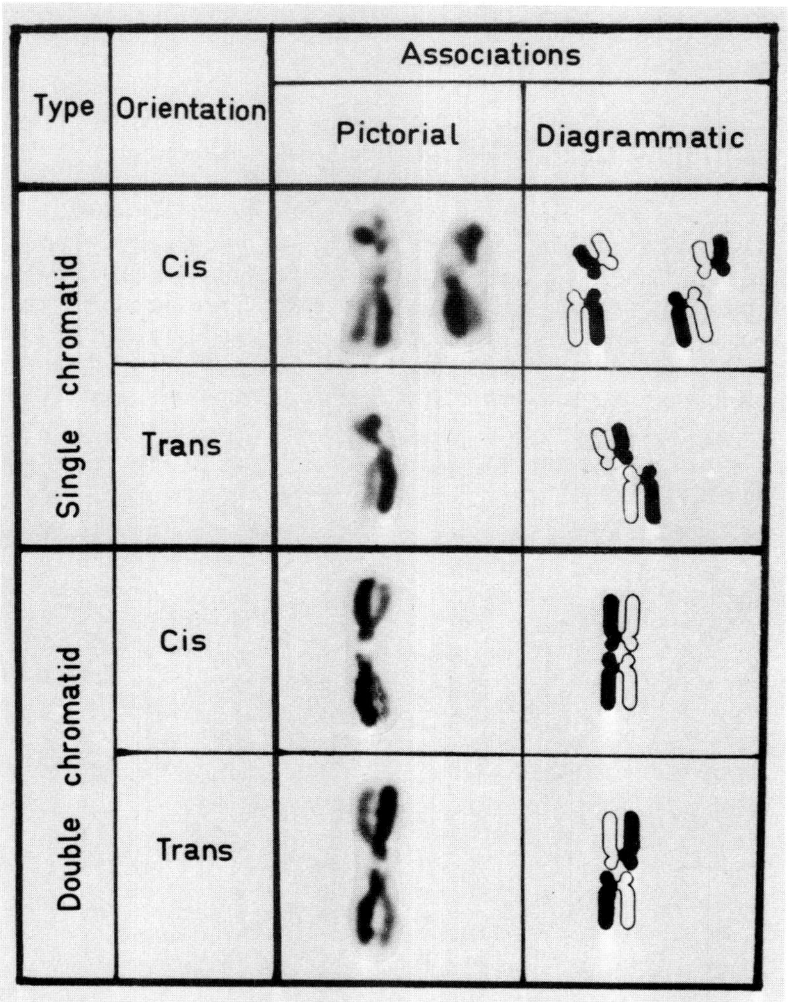

FIG. 5. Pictorial demonstration of cis (between two darkly stained chromatids, i.e., unifiliarly BudR incorporated; between two lightly stained chromatids, i.e., between bifiliarly BudR incorporated) and trans (between the light and dark chromatids) associations.

matid in "second" and "third" generation cycles (Chemitiganti *et al.*, 1984) (Fig. 5).

The frequency of acrocentric associations is influenced by various physiological and physical factors, e.g., cell cycle duration and NORs activity (Mattevi and Salzano, 1975; Sigmund *et al.*, 1979), level of thyroid hormones (Hansson, 1971; Nilsson *et al.*, 1975; Zankl *et al.*, 1980), aging (Cooke, 1972; Mattei *et al.*, 1976; Liem *et al.*, 1977; Kadotani *et al.*, 1978), and culture and technical conditions (Zang and Back, 1969; Nankin, 1970; Hansson, 1970; Verma and Dosik, 1979).

V. Role of NORs in Chromosomal Alterations

The chromosomal associations resulted from NORs play an important role in at least three types of chromosomal disorders. The most frequent of them is the meiotic nondisjunction causing the trisomic condition (Ferguson-Smith and Handmaker, 1961; Ohno *et al.*, 1961) for which the classical examples are Down and Patau syndromes (Cook, 1971, 1972; Rosenkraz and Holzer, 1972; Capoa *et al.*, 1973; Curtis, 1974; Nakagome, 1973). About 40% of lethal trisomies in human is due to the acrocentric NOR bearing chromosomes (Boue and Boue,

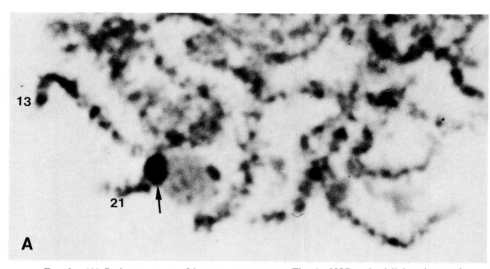

FIG. 6.   (A) Pachytene stage of human spermatocyte. The Ag-NOR stained light micrograph showing nucleolus connected with two acrocentric bivalents identified as 13 and 21 on the basis of their chromomere pattern. (B) Pachytene oocyte from an 18-week human fetus; the segregated fibrillar center (FC) of the nucleolus is connected with intermingled chromatin fibers which emanate from two acrocentric bivalents. SY, synaptonemal complex. (Courtesy of Dr. A. Stahl.)

1977) and the majority of them result from maternal meiotic nondisjunction (Jacobs and Morton, 1977; Mattei *et al.*, 1979). The recent estimates made by Verma *et al.* (1985) support the above studies with 78 and 22% incidence of Down's syndrome due to maternal and paternal nondisjunction, respectively, and

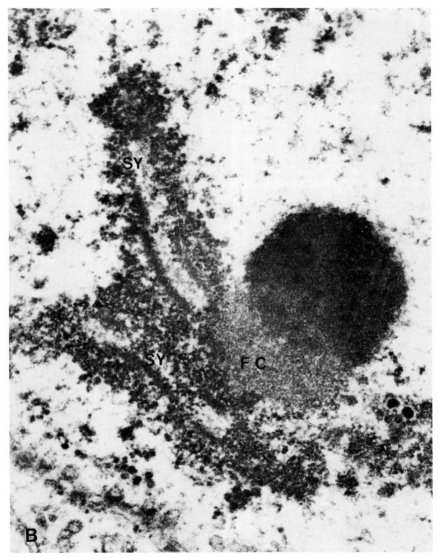

FIG. 6B.

with higher frequency at M I (78.7%) than at M II (21.3%). Evidently both paternal and maternal origin of trisomy are due to meiotic nondisjunction resulting from close association of rDNA cistrons of different NORs in fibrillar centers during pachytene stages (Fig. 6). Probably, the individuals with highly active NORs in chromosome 21 may have relatively higher risk of meiotic nondisjunc-

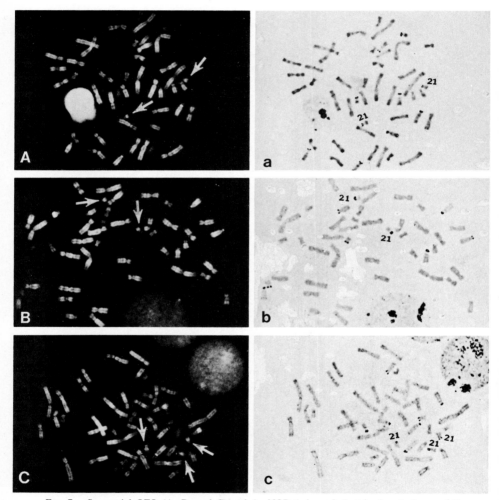

FIG. 7.   Sequential QFQ (A, B, and C) and Ag-NOR (a,b, and c) stained metaphases of the mother (A), father (B), and the proband (C) with trisomy 21. The chromosome 21s are identified with arrows in QFQ and with numbers in Ag-NOR metaphases. The origin of extra chromosome 21 was found to be paternal. The nondisjunction in the father was due to active NORs in chromosome 21s (Verma *et al.*, 1983).

tion (Fig. 7). However, the differences in the ratio of parental origin are probably associated with functional and structural diversity of NORs during spermatogenesis and oogenesis. The male meiotic nondisjunction is relatively less frequent probably due to the absence of NOR expression during the crutial stages of disjunction, i.e., diakinesis and metaphase II in spermatogenesis (Schmid *et al.*, 1977, 1983; Schwarzacher *et al.*, 1978; Ironside and Faed, 1979) while more the complex nature of nucleolar expression in oogenesis with multiple micronucleoli (Stahl *et al.*, 1975; Mirre *et al.*, 1980; Hartung *et al.*, 1983), additional rDNA (Wolgemuth-Jara Show *et al.*, 1977; Wolgemuth *et al.*, 1979), also the structural rearrangements during prophase I (Hartung *et al.*, 1983), promote the possibility of nondisjunction. Further, the comparative study with human and mouse tissues reveals the significant relation between structural patterns of NORs with different degrees of nondisjunction (Mirre *et al.*, 1980).

The second type of chromosomal changes, Robertsonian type translocations between acrocentric chromosomes in human karyotype is related to formation of nucleoli and NOR associations (Ohno *et al.*, 1961). Recently, this phenomenon is further evidenced with studies on mouse NOR bearing chromosomes participate at significantly higher rate than the other counterparts (O. J. Miller *et al.*, 1978). In humans the translocations occur between homologous (Buys *et al.*,

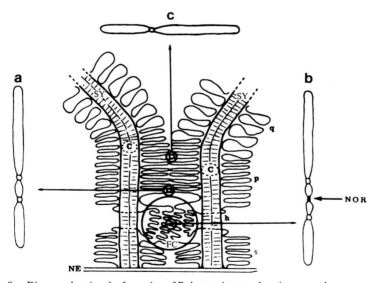

FIG. 8. Diagram showing the formation of Robertsonian translocations at pachytene stage after a breakage/reunion involving nonhomologous chromatids. (a) Dicentric without NOR; (b) dicentric with NOR; (c) centromere; SY, synaptonemal complex; NE, nuclear envelop; FC, fibrillar center of the nucleolus; p, proximal part of the short arm; q, long arm; h, secondary constriction region; s, satellite. (Courtesy of Dr. A. Stahl.)

1978; Zankl and Hahmann, 1978) or heterologous pairs (Archidiacono et al., 1977; Hurley and Pathak, 1977; Gosden et al., 1978; Mattei et al., 1979; Mikkelsen et al., 1980) involving the entire chromosomes along with NOR bearing short arms or with centromeric fusion eliminating NOR. The later type of translocations seems to be much more frequent (Mattei et al., 1979; Mikkelsen et al., 1980). The possible role of NOR association and mechanism involved in production of de novo translocations during human gametogenesis was reported by Stahl et al. (1983) (Fig. 8). Among the altered chromosomes, only those retaining the NORs participate in nucleolar associations (Schmid et al., 1974; Orye, 1974; D. A. Miller et al., 1977; Zankl and Hahmann, 1978; Mattei et al., 1979). Yet another type of abnormality resulting in the alteration of a number of NOR sites is the bisatellited supernumerary chromosomes in normal and abnormal phenotypes (Babu et al., 1983) (for review see Babu et al., 1985).

Biochemical studies on trisomy 21 cases have shown the increased rDNA while it is reduced in translocations with the loss of NORs (Bross et al., 1973) compared to normal probands. The compensatory mechanism for the lost NORs was suggested (Hansson, 1975; Gosden et al., 1978; Zankl and Hahmann, 1978; Zankl and Zang, 1979) though no such relation could be found by Nikolis et al. (1981).

VI. NORs and Nucleoli

The ribosomal genes present in NORs are located in the fibrillar centers (see Gossens and Lepoint, 1979, for nomenclature) of organized nucleoli (Busch and Smetana, 1970; Recher et al., 1976). The transcriptional activity of ribosomal DNA cistrons takes place in the peripheral parts, i.e., at the border of dense fibrillar component (e.g., Bouteille et al., 1974, 1982; Goessens, 1979; Mirre and Stahl, 1978, 1981; Fakan and Puvion, 1980) and primary transcriptional products are found in the fibrillar component while the processed products are located in the granular component of nucleolus. Silver methods stain mainly the fibrillar centers and to an extent the fibrillar components (e.g., Wachtler and Musil, 1979; Bourgeois et al., 1979; Goessens, 1979; Ellinger and Wachtler, 1980; Hernandez-Verdun et al., 1980; Hernandez-Verdun, 1983; Goessens and Lepoint, 1982; Mirre and Knibiehler, 1982) and is similar to the Ag-NOR staining in chromosomes. Further, Angelier et al. (1982) have demonstrated the presence of silver-positive proteins in the transcribed part of nucleolar gene while they are absent in untranscribed spacer regions in molecular spreads (Fig. 2). However, the special relationship between fibrillar center and the transcriptional sites is unclear. The degree of NOR activity is represented by different types of nucleoli viz. small ring-shaped nucleoli with low activity and compact nucleoli with full activity. The nucleoli with nucleolonema is the transitional stage between the two forms (review by Schwarzacher and Wachtler, 1983). The NOR activity and type of nucleoli depend on the cellular functional activity, and both of them change with altered cell function as seen in phytohemagglutinin-stimulated lymphocytes (Arrighi *et al.*, 1980; Wachtler *et al.*, 1980, 1982).

VII. Amplification and Regulation of Human rDNA

The amount of rDNA is highly variable from chromosome to chromosome and from individual to individual (Warburton *et al.*, 1976; D. A. Miller *et al.*, 1978). These variations probably result from the unequal crossing over or duplication. The human chromosomes 14p+ reported by D. A. Miller *et al.* (1978) and Lau *et al.* (1979) alone are approximated to consist of 6–8 times excessive copies of rDNA than the normal single chromosome. However, only a fraction of this amplified rDNA is transcriptionally active as demonstrated by silver staining. The importance of 5-methyl cytosine (5 MeC) in human rDNA regulation has been emphasized by Tantravahi *et al.* (1981) because of correlation between high concentration of 5MeC in the inactive NORs (also see O. J. Miller *et al.*, 1981, for other systems). The molecular aspects of rDNA regulation and amplification have been reviewed extensively (O. L. Miller, 1981; Perry, 1981; MacGregor, 1982; Cullis, 1982; Moss and Birnstiel, 1982).

VIII. NOR Heteromorphisms

The short arms of human acrocentric chromosomes are the sites of heteromorphisms detectable both by conventional staining (Court-Brown and Jacobs, 1965) and by various banding techniques. There are at least three bands, which have been recognized as variable by the Paris Conference (1971, 1975). In particular by QFQ (Q-bands by fluorescence using quinacrine; ISCN 1978) and CBG (C-bands by barium hydroxide using Giemsa) techniques, bands p.11, p.12, and p.13 exhibit heteromorphisms (Caspersson *et al.*, 1971; McKenzie and Lubs, 1975; Verma and Dosik, 1977, 1980). Employing the RFA technique, all the three bands, i.e., p.11, p.12, and p.13 show heteromorphisms (Verma and Lubs, 1975; Verma *et al.*, 1977a–c). While using the Ag-NOR technique only band p.12, i.e., the nucleolar organizer region is darkly stained (Vedbrat *et al.*, 1980). It is a documented fact that each individual has characteristic pattern and model number of Ag-stainable NORs which are inherited from generation to generation (Varley, 1977; Mikelsaar *et al.*, 1977a,b; Markovic *et al.*, 1978; Mikelsaar and Schwarzacher, 1978; Mikelsaar and Ilus, 1979; Ray and Pearson, 1979; Sofuni *et al.*, 1980; Taylor and Deleon, 1981). The darkly stained NORs using the Ag-NOR technique are equivalent to the pale green color using the

RFA technique (Verma *et al.*, 1983) and this color is not seen in any other part of the human genome. However, the number of acrocentric chromosomes showing green color varies from person to person but is consistent within the same person (Verma and Lubs, 1976a,b; Verma and Dosik, 1981a). The morphological features of NORs using two different techniques have further been correlated. It is concluded that there is no direct relationship between a heteromorphism of NORs identified by RFA technique (Fig. 9). Further, it is emphasized that the ex-

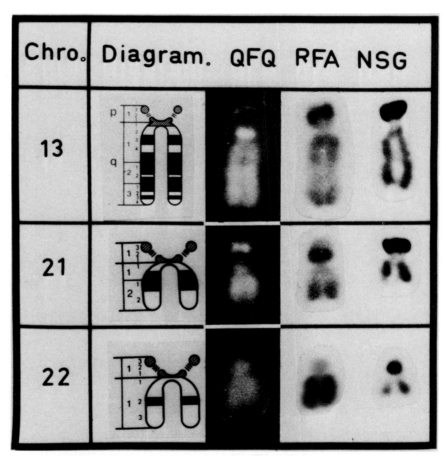

FIG. 9.   Demonstration of nucleolar organizer regions (NORs) in chromosomes 13, 21, and 22 by QFQ, RFA, and Ag-NOR banding techniques. A very small satellite was seen on chromosome 13 by QFQ technique. Satellites with brilliant QFQ intensity stain deep red by RFA and do not show well in black and white when they are printed from color transparency film. However, a pale green color was not seen in chromosome 22, yet NOR of this chromosome was found to have silver deposits by the Ag-NOR technique suggesting differential staining properties of NORs by various staining methods.

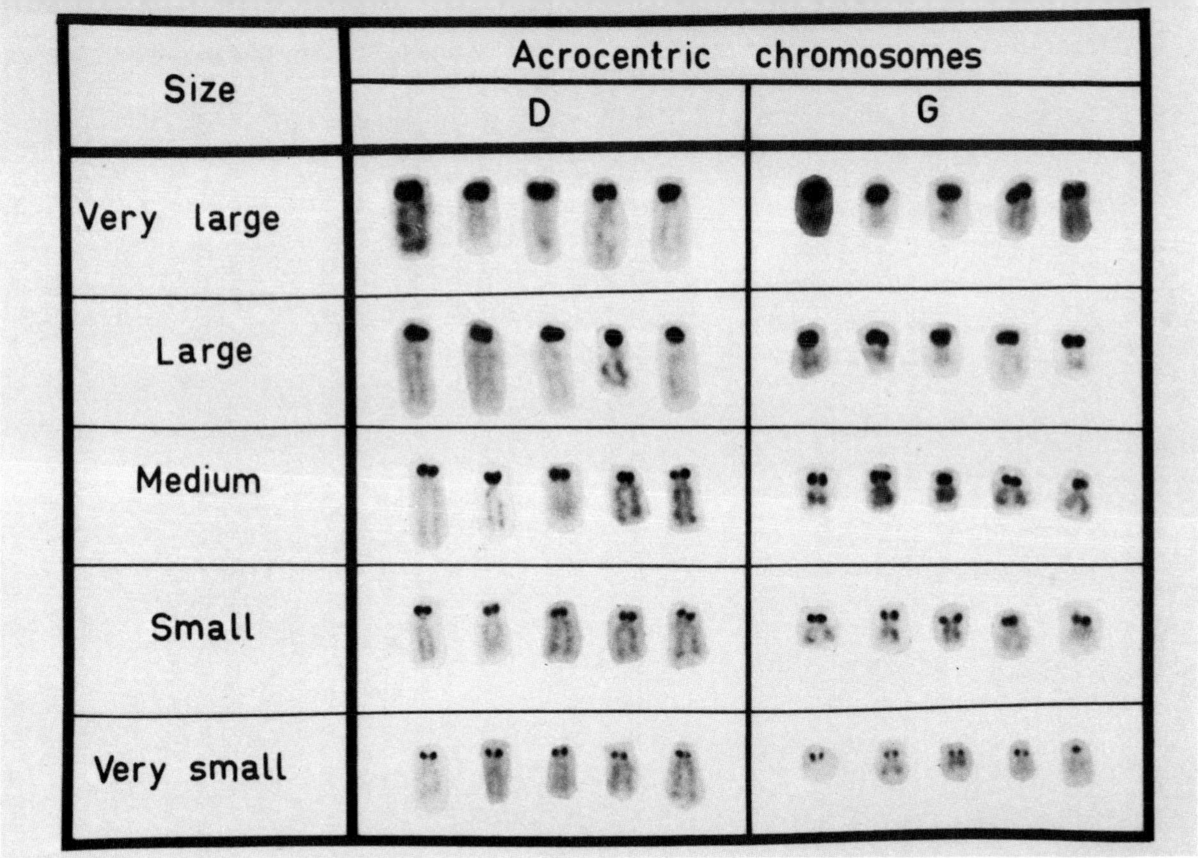

FIG. 10. Demonstration of different sizes of NORs by Ag-NOR technique. Size variation with the group is continuous and rather discrete (see also Verma *et al.*, 1981).

pression of NORs was more precise by Ag staining than by reverse banding (Verma et al., 1981). Based on silver deposit, we have made an attempt to classify these regions into 5 sizes (Fig. 10). At the cytological level these large NORs presumably contain multiple copies while small NORs have fewer compies of the 18 S and 28 S ribosomal cistrons. Further, unusually large sized short arms with duplicate and triplicate NORs and/or satellite have been reported (Archidiacono et al., 1977; Balicek and Zizka, 1980; Sofuni et al., 1980) with extreme examples of 14p+ (D. A. Miller et al., 1978; Lau et al., 1979) and 22p+ (Bernstein et al., 1981).

Also, the polymorphic forms of different acrocentrics have been evaluated using base specific fluorochromes, antibiotics, and fluorescence antibodies (Schnedl, 1978; Okamoto et al., 1981). Racial differences in acrocentric short arms and Ag-NOR activity were observed by Mikelsaar and Ilus (1979), Verma et al. (1981) and Verma and Dosik (1981a,b).

IX. Clinical Implications

The human acrocentric chromosomes due to the presence of NORs on short arms have a tendency to remain associated with each other during the cell division which have deletereous consequences through meiotic nondisjunction. Although the expression and clinical implications for all acrocentrics are unknown, the priliminary studies suggest that trisomy 21 (Down's syndrome) is caused by expression of NORs in chromosome 21 (Verma et al., 1983) (Fig. 7).

A large number of cases with various types of small supernumerary (accessary) chromosomes in normal as well as in abnormal children have been noted (Fig. 1D), which were found to have Ag-positive NORs. The prognosis of such cases is quite obscure. However, from the available reports, the individuals with accessary chromosomes with positive NORs have a better prognosis as they are clinically more normal than those with negative NORs (Babu et al., 1985).

Besides accessary chromosomes there is another variant chromosome 17 which bears an NOR site in normal as well as in clinically abnormal human populations. Using silver staining it has been now proven that such chromosomes have no NOR sites and does not originate from acrocentrics. Therefore, it is considered a rare heteromorphism (Verma et al., 1983; Verma and Dosik, 1979).

With the use of silver staining, the exact break points have been localized in Robertsonian translocations (Mattei et al., 1979) and the extent of the trisomic nature has further been explored in the human population (Mikkelsen et al., 1980). The variation in the activity of nucleolar organizer regions in different tissues has been demonstrated in normal and leukemic cells and contrary findings exist (Zankl et al., 1980, 1982; Trent et al., 1981; Reeves et al., 1982).

Very recently, Verma et al. (1981) have provided the quantitative data on the different sizes of NORs in the human population (Fig. 1). The biological and clinical implications of NOR size heteromorphisms of human acrocentrics are poorly understood. Racial differences in the expression of NORs are suggested (Verma *et al.*, 1981). Racial differences have anthropological interest and have great value in linkage and population studies and in prenatal diagnosis.

REFERENCES

Angelier, N., Hernandez-Verdun, D., and Bouteille, M. (1982). *Chromosoma* **86**, 661.
Archidiacono, N., Capoa, A., de, Ferraro, M., Pelliccia, F., Rocchi, A., and Rocchi, M. (1977). *Hum. Genet.* **37**, 285.
Ardito, G., Lamberti, L., and Brogger, A. (1976). *Ann. Hum. Genet.* **41**, 455.
Arrighi, F. E., Lau, Y. F., and Spallone, A. (1980). *Cytogenet. Cell Genet.* **26**, 244.
Babu, K. A., Verma, R. S., Rodriguez, J., Rosenfeld, W., and Dosik, H. (1983). *J. Cell Biol.* **97**, 380a.
Babu, K. A., Verma, R. S., Rodriguez, Rosenfeld, W., and Dosik, H. (1985). Submitted.
Balicek, P., and Zizka, J. (1980). *Hum. Genet.* **54**, 343.
Beek, B. (1981). *Hum. Genet.* **59**, 240.
Bernstein, R., Dawson, B., and Griffiths, J. (1981). *Hum. Genet.* **58**, 135.
Bloom, S. E., and Goodpasture, C. (1976). *Hum. Genet.* **34**, 199.
Bobrow, M., and Heritage, J. (1980). *Nature (London)* **288**, 79.
Boue, A., and Boue, J. (1977). *J. Gynecol. Obstet. Biol. Reprod.* **6**, 5.
Bourgeois, C. A., Hernandez-Verdun, D., Hubert, J., and Bouteille, M. (1979). *Exp. Cell Res.* **123**, 449.
Bouteille, M., Laval, M., and Dupuy-Coin, A. M. (1974). *In* "The Cell Nucleus" (H. Busch, ed.), pp. 5–64. Academic Press, New York.
Bouteille, M., Hernandez-Verdun, D., Dupuy-Coin, A. M., and Bourgeois, C. A. (1982). *In* "The Nucleolus" (G. Jordan and C. A. Cullis, eds.), pp. 179–211. Cambridge Univ. Press, London and New York.
Bross, K., and Krone, W. (1972). *Hum. Genet.* **14**, 137.
Bross, K., Dittes, H., Krone, W., Schmid, M., and Vogel, W. (1973). *Humangenetik* **20**, 223.
Busch, H., and Smetana, K. (1970). *In* "The Cell Nucleus" (H. Busch, ed.). Academic Press, New York.
Busch, H., Lischwe, M. A., Michalik, J., Pui Kwong, C. H., and Busch, H. K. (1982). *In* "The Nucleolus" (G. Jordan and C. A. Cullis, eds.), pp. 43–71. Cambridge Univ. Press, London and New York.
Buys, C. H. C. M., and Osinga, J. (1980). *Chromosoma* **77**, 1.
Buys, C. H. C. M., Osinga, J., Gouw, W. L., and Anders, G. J. P. A. (1978). *Hum. Genet.* **44**, 173.
Buys, C. H. C. M., Osinga, J., and Anders, G. J. P. A. (1979). *Mech. Ageing Dev.* **11**, 55.
Capoa, A., de, Rocchi, A., and Gigliani, F. (1973). *Humangenetik* **18**, 111.
Capoa, A., de, Ferraro, M., Mendez, F., Mostacci, C., Pelliccia, F., and Rocchi, A. (1978). *Hum. Genet.* **44**, 71.
Caspersson, T., Lomakka, G., and Zech, L. (1971). *Heriditas* **67**, 89.
Chemitiganti, S., Verma, R. S., Vedbrat, S., and Dosik, H. (1984). *Can. J. Genet. Cytol.* **26**, 137.
Cheng, D. M., Denton, T. E., Liem, S. L., and Elliot, C. L. (1981). *Clin. Genet.* **19**, 145.
Chicago Conference (1966). *Birth Defects, Orig. Art. Ser.* **2** (2).

Cohen, M. M., and Shaw, M. W. (1967). *Ann. Hum. Genet.* **31**, 129.

Cooke, P. (1971). *Humangenetik* **13**, 309.

Cooke, P. (1972). *Humangenetik* **17**, 29.

Court-Brown, W. M., and Jacobs, P. A. (1965). *Lancet* **2**, 561.

Cullis, C. A. (1982). In "The Nucleolus" (G. Jordan and C. A. Cullis, eds.), pp. 103–112. Cambridge Univ. Press, London and New York.

Curtis, D. J. (1974). *Humangenetik* **22**, 17.

Denton, T. E., Liem, S. L., Cheng, K. M., and Barrett, J. V. (1981). *Mech. Ageing Dev.* **15**, 1.

Dittes, H., Krone, W., Bross, K., Schmid, M., and Vogel, W. (1975). *Hum. Genet.* **26**, 47.

Egolina, N. A., Davudov, A. Z., Benjusch, V. A., and Zakharov, A. F. (1981). *Biul. Eksp. Biol. Med.* **91**, 349.

Eliceiri, G. L., and Green, H. (1969). *J. Mol. Biol.* **41**, 253.

Ellinger, A., and Wachtler, F. (1980). *Mikroskopie* **36**, 330.

Evans, H. J., Buckland, R. A., and Pardue, M. L. (1974). *Chromosoma* **48**, 405.

Fakan, S., and Puvion, E. (1980). *Int. Rev. Cytol.* **65**, 255.

Faust, J., and Vogel, W. (1974). *Nature (London)* **249**, 352.

Ferguson-Smith, M. A. (1964). *Cytogenetics* **3**, 124.

Ferguson-Smith, M. A., and Handmaker, S. D. (1961). *Lancet* **1**, 638.

Ferraro, M., Archidiacono, N., Pellicia, F., Rocchi, M., Rocchi, A., and Capoa, A., de (1977). *Exp. Cell Res.* **104**, 428.

Funaki, K., Matsui, S., and Sasaki, M. (1975). *Chromosoma* **49**, 357.

Gall, J. G., and Pardue, M. L. (1969). *Proc. Natl. Acad. Sci. U.S.A.* **63**, 378.

Galperin-Lemaitre, H., Hens, L., and Sele, B. (1980). *Hum. Genet.* **54**, 349.

Gerlach, W. L. (1977). *Chromosoma* **62**, 49.

Goessens, G. (1976). *Exp. Cell Res.* **100**, 88.

Goessens, G. (1979). *Cell Tissue Res.* **200**, 159.

Goessens, G., and Lepoint, A. (1979). *Biol. Cell.* **35**, 211.

Goessens, G., and Lepoint, A. (1982). *Biol. Cell.* **43**, 139.

Goodpasture, C., and Bloom, S. E. (1975). *Chromosoma* **53**, 37.

Goodpasture, C., Bloom, S. E., Hsu, T. C., and Arrighi, F. E. (1976). *Am. J. Hum. Genet.* **28**, 559.

Gosden, J. R., Gosden, C., Lawrie, S. S., and Mitchell, A. R. (1978). *Hum. Genet.* **41**, 131.

Gosden, J. R., Gosden, C. M., Lawrie, S. S., and Buckton, K. E. (1979). *Clin. Genet.* **15**, 518.

Hansson, A. (1970). *Hereditas* **66**, 31.

Hansson, A. (1971). *Hereditas* **69**, 269.

Hansson, A. (1975). *Hereditas* **81**, 101.

Hansson, A. (1979). *Hereditas* **90**, 59.

Hartung, M., Keeling, J. W., Patel, C., Bobrow, M., and Stahl, A. (1983). *Cytogenet. Cell Genet.* **35**, 2.

Hatlen, L., and Attardi, G. (1971). *J. Mol. Biol.* **56**, 535.

Hayata, I., Oshimura, M., and Sandberg, A. A. (1977). *Hum. Genet.* **36**, 55.

Henderson, A. S., and Atwood, K. C. (1976). *Hum. Genet.* **31**, 113.

Henderson, A. S., Warburton, D., and Atwood, K. C. (1972). *Proc. Natl. Acad. Sci. U.S.A.* **69**, 3394.

Henderson, A. S., Warburton, D., and Atwood, K. C. (1973). *Nature (London)* **245**, 95.

Henderson, A. S., Warburton, D., and Atwood, K. C. (1974a). *Chromosoma* **44**, 367.

Henderson, A. S., Warburton, D., and Atwood, K. C. (1974b). *Chromosoma* **46**, 433.

Hernandez-Verdun, D. (1983). *Biol. Cell* **49**, 191.

Hernandez-Verdun, D., and Bouteille, M. (1979). *J. Ultrastruct. Res.* **69**, 164.

Hernandez-Verdun, D., Hubert, J., Bourgeois, C. A., and Bouteille, M. (1980). *Chromosoma* **79**, 349.

Hernandez-Verdun, D., Derenzini, M., and Bouteille, M. (1982). *Chromosoma* **85,** 461.
Hilwig, I., and Gropp, A. (1973). *Exp. Cell Res.* **81,** 474.
Houghton, J. A., and Houghton, S. E. (1978). *Sci. Prog.* **65,** 331.
Howell, W. M., and Black, D. A. (1978). *Hum. Genet.* **43,** 53.
Howell, W. M., and Black, D. A. (1980). *Experientia* **36,** 1014.
Howell, W. M., and Denton, T. E. (1974). *Experientia* **30,** 1364.
Howell, W. M., and Hsu, T. C. (1979). *Chromosoma* **73,** 61.
Howell, W. M., Denton, T. E., and Diamond, J. R. (1975). *Experientia* **31,** 260.
Hsu, T. C. (1973). *Annu. Rev. Cytol.* **7,** 153.
Hsu, T. C., Spirito, S. E., and Pardue, M. L. (1975). *Chromosoma* **53,** 25.
Hubbell, H. R., and Hsu, T. C. (1977). *Cytogenet. Cell Genet.* **19,** 185.
Hubbell, H. R., Rothblum, L. I., and Hsu, T. C. (1979). *Cell Biol. Int. Rep.* **3,** 615.
Huberman, J. A., and Attardi, G. (1967). *J. Mol. Biol.* **29,** 487.
Hurley, J. E., and Pathak, S. (1977). *Hum. Genet.* **35,** 169.
Ironside, J., and Faed, M. J. W. (1979). *Hum. Genet.* **48,** 39.
ISCN (1978). An International System for Human Cytogenetic Nomenclature. *Birth Defects* **XIV,** 47.
Jacobs, P. A., and Morton, N. M. (1977). *Hum. Genet.* **27,** 59.
Jacobs, P. A., Mayer, M., and Morton, N. E. (1976). *Am. J. Hum. Genet.* **28,** 567.
Jeanteur, P., and Attardi, G. (1969). *J. Mol. Biol.* **45,** 305.
Jewell, D. C. (1979). *Chromosoma* **71,** 129.
Jewell, D. C. (1981). *Stain. Tech.* **56,** 427.
Kadotani, T., Watanabe, Y., and Makino, S. (1978). *Proc. Jpn. Acad.* **54,** 277.
Kling, H., Lepre, L., Krone, W., Olert, J., and Sawatzki, G. (1980). *Experientia* **36,** 249.
Kodama, Y., Yoshida, M. C., and Sasaki, M. (1980). *Jpn. J. Hum. Genet.* **25,** 229.
Lau, Y.-F., and Arrighi, F. E. (1977). *In* Monograph concerning Seminar-workshop "Aspects of the Chromosome Organization and Function" (M. E. Drets, N. Brum-Zorilla, and G. A. Folle, eds.), pp. 49–55, Montevideo, Uruguay.
Lau, Y.-F., Pfeiffer, R. A., Arrighi, F. E., and Hsu, T. C. (1978). *Am. J. Hum. Genet.* **30,** 76.
Lau, Y.-F., Wertelecki, W., Pfeiffer, R. A., and Arrighi, F. E. (1979). *Hum. Genet.* **46,** 75.
Liem, S. L., Denton, T. E., and Cheng, K. M. (1977). *Clin. Genet.* **12,** 104.
Lischwe, M. A., Smetana, K., Olson, M. O. J., and Busch, H. (1979). *Life Sci.* **25,** 701.
McClintock, B. (1934). *Z. Zellforsch. Mikrosk. Anat.* **21,** 294.
MacGregor, H. C. (1982). *In* "The Nucleolus" (E. G. Jordan and C. A. Cullis, eds.), pp. 129–151, Cambridge Univ. Press, London and New York.
McKenzie, W. H., and Lubs, H. A. (1975). *Cytogenet. Cell Genet.* **14,** 97.
Markovic, V. D., Worton, R. G., and Berg, J. M. (1978). *Hum. Genet.* **41,** 181.
Matsui, S. (1974). *Exp. Cell Res.* **88,** 88.
Matsui, S., and Sasaki, M. (1973). *Nature (London)* **246,** 148.
Mattei, J. F., Ayme, S., Mattei, M. G., Gonvernet, J., and Girand, F. (1976). *Hum. Genet.* **34,** 185.
Mattei, M. G., Mattei, J. F., Ayme, S., and Giraud, F. (1979). *Hum. Genet.* **50,** 53.
Mattevi, M. S., and Salzano, F. M. (1975). *Humangenetik* **29,** 265.
Mikelsaar, A. V., and Ilus, T. (1979). *Hum. Genet.* **51,** 281.
Mikelsaar, A. V., and Schwarzacher, H. G. (1978). *Hum. Genet.* **42,** 291.
Mikelsaar, A. V., Schmid, M., Krone, W., Schwarzacher, H. G., and Schnedl, W. (1977a). *Hum. Genet.* **37,** 73.
Mikelsaar, A. V., Schwarzacher, H. G., Schnedl, W., and Wagenbichler, P. (1977b). *Hum. Genet.* **38,** 183.
Mikkelsen, M., Basli, A., and Poulsen, H. (1980). *Cytogenet. Cell Genet.* **26,** 14.
Miller, D. A., Dev, V. G., Tantravahi, R., and Miller, O. J. (1976). *Exp. Cell Res.* **101,** 235.

Miller, D. A., Tantravahi, R., Dev, V. G., and Miller, O. J. (1977). *Am. J. Hum. Genet.* **29,** 490.
Miller, D. A., Berg, W. R., Warburton, D., Dev, V. G., and Miller, O. J. (1978). *Hum. Genet.* **43,** 289.
Miller, O. J., Miller, D. A., Dev, V. G., Tantravahi, R., and Groce, C. M. (1976). *Proc. Natl. Acad. Sci. U.S.A.* **73,** 4531.
Miller, O. J., Miller, D. A., Tantravahi, R., and Dev, V. G. (1978). *Cytogenet. Cell Genet.* **20,** 40.
Miller, O. J., Tantravahi, U., Katz, R., Erlanger, B. F., and Guntaka, R. V. (1981). *In* "Genes, Chromosomes and Neoplasia" (F. E. Arrighi, P. N. Rao, and E. Stubblefield, eds.), pp. 253–270. Raven, New York.
Miller, O. L., Jr. (1981). *J. Cell Biol.* **91,** 15s.
Mirre, C., and Knibiehler, B. (1981). *Biol. Cell.* **42,** 73.
Mirre, C., and Knibiehler, B. (1982). *J. Cell Biol.* **55,** 247.
Mirre, C., and Stahl, A. (1978). *J. Ultrastruct. Res.* **64,** 377.
Mirre, C., and Stahl, A. (1981). *J. Cell Sci.* **48,** 105.
Mirre, C., Hartung, M., and Stahl, A. (1980). *Proc. Natl. Acad. Sci. U.S.A.* **77,** 6017.
Moss, T., and Birnstiel, M. (1982). *In* "The Nucleolus" (E. G. Jordan and C. A. Cullis, eds.), pp. 73–85. Cambridge Univ. Press, London and New York.
Nakagome, Y. (1973). *Cytogenet. Cell Genet.* **12,** 336.
Nankin, H. R. (1970). *Cytogenetics* **9,** 42.
Nielsen, K., Marcus, M., and Gropp, A. (1979). *Hereditas* **90,** 31.
Nikolis, J., Kekic, V., and Diklic, V. (1981). *Hum. Genet.* **59,** 342.
Nilsson, C., Hansson, A., and Nilsson, G. (1975). *Hereditas* **80,** 157.
Ohno, S., Trujillo, J. M., Kaplan, W. D., and Kinosita, R. (1961). *Lancet* **2,** 123.
Okamoto, E., Miller, D. A., Erlanger, B. F., and Miller, O. J. (1981). *Hum. Genet.* **58,** 255.
Olert, J., Sawatzki, G., Kling, H., and Gebauer, T. (1979). *Histochemistry* **60,** 91.
Orye, E. (1974). *Humangenetik* **22,** 299.
Pardue, M. L., and Birnstiel, M. L. (1973). *Symp. Med. Hoechst* **6,** 75–85.
Pardue, M. L., and Gall, J. G. (1969). *Proc. Natl. Acad. Sci. U.S.A.* **64,** 600.
Pardue, M. L., and Hsu, T. C. (1975). *J. Cell Biol.* **64,** 251.
Pardue, M. L., Brown, D. D., and Birnstiel, M. L. (1973). *Chromosoma* **42,** 191.
Paris Conference (1971). *Birth Defects* **VIII.**
Paris Conference (1975). *Birth Defects Suppl.* **XI.**
Patil, S. R., and Lubs, M. A. (1971). *Humangenetik* **13,** 157.
Perry, R. P. (1981). *J. Cell Biol.* **91,** 28s.
Pimpinelli, S., Santini, G., and Gatti, M. (1976). *Chromosoma* **57,** 377.
Ray, M., and Pearson, J. (1979). *Hum. Genet.* **48,** 201.
Recher, L., Skyes, J. A., and Chan, H. (1976). *J. Ultrastruct. Res.* **56,** 152.
Reeves, B. R., Casey, G., and Harris, H. (1982). *Cancer Genet. Cytogenet.* **6,** 223.
Robbins, S. L., and Angell, M. (1976). "Basic Pathology," 2nd Ed. Saunders, Philadelphia, Pennsylvania.
Rosenkrauz, W., and Holzer, S. (1972). *Humangenetik* **16,** 147.
Rothfels, K. H., and Siminovitch, L. (1958). *Chromosoma* **9,** 163.
Schmid, M., Krone, W., and Vogel, W. (1974). *Hum. Genet.* **23,** 267.
Schmid, M., Hofgartner, F. J., Zenzes, M. T., and Engel, W. (1977). *Hum. Genet.* **38,** 279.
Schmid, M., Muller, H., Stasch, S., and Engel, W. (1983). *Hum. Genet.* **64,** 363.
Schnedl, W. (1978). *Hum. Genet.* **41,** 1.
Schwarzacher, H. G., and Watchler, F. (1983). *Hum. Genet.* **63,** 89.
Schwarzacher, H. G., Mikelsaar, A. V., and Schnedl, W. (1978). *Cytogenet. Cell Genet.* **20,** 24.
Sigmund, J., Schwarzacher, H. G., and Mikelsaar, A. V. (1979). *Hum. Genet.* **50,** 81.
Sinclair, J. H., and Brown, D. D. (1971). *Biochemistry* **10,** 2761.
Sofuni, T., Tanabe, K., and Awa, A. A. (1980). *Hum. Genet.* **55,** 265.

Stack, S. M. (1974). *Chromosoma* **47**, 361.

Stahl, A. (1982). *In* "The Nucleolus", (E. G. Jordan and C. A. Cullis, eds.), pp. 1–24. Cambridge Univ. Press, London and New York.

Stahl, A., Luciani, J. M., Devictor, M., Capodano, A. M., and Gagne, R. (1975). *Humangenetik* **26**, 315.

Stahl, A., Luciani, J. M., Hartung, M., Devictor, M., Berge-Lefrance, J. L., and Guichaoua, M. (1983). *Proc. Natl. Acad. Sci. U.S.A.* **80**, 5946.

Strehler, B. L., and Chang, Mei-ping (1979). *Mech. Ageing Dev.* **11**, 379.

Strehler, B. L., Chang, Mei-ping, and Johnson, L. K. (1979). *Mech. Ageing Dev.* **11**, 371.

Strobel, R. J., Pathak, S., and Hsu, T. C. (1981). *Hum. Genet.* **59**, 259.

Tantravahi, R., Miller, D. A., Dev, V. G., and Miller, O. J. (1976). *Chromosoma* **56**, 15.

Tantravahi, R., Miller, D. A., and Miller, O. J. (1977). *Cytogenet. Cell Genet.* **18**, 364.

Tantravahi, U., Berg, W. R., Wertelecki, W., Erlanger, B. E., and Miller, O. J. (1981). *Hum. Genet.* **56**, 315.

Taylor, E. F., and Deleon, P. A. M. (1980). *Hum. Genet.* **54**, 217.

Taylor, E. F., and Deleon, P. A. M. (1981). *Am. J. Hum. Genet.* **33**, 67.

Trent, J. M., Carlin, D. A., and Davis, J. R. (1981). *Cytogenet. Cell Genet.* **30**, 31.

Varley, J. M. (1977). *Chromosoma* **61**, 207.

Vedbrat, S., Verma, R. S., and Dosik, H. (1979). *Stain. Tech.* **54**, 107.

Vedbrat, S., Verma, R. S., and Dosik, H. (1980). *Stain. Tech.* **55**, 77.

Verma, R. S., and Babu, K. A. (1984). *Karyogram* **10**, 4.

Verma, R. S., and Dosik, H. (1977). *IRCS Med. Sci.* **5**, 535.

Verma, R. S., and Dosik, H. (1979). *Clin. Genet.* **15**, 450.

Verma, R. S., and Dosik, H. (1980). *Int. Rev. Cytol.* **62**, 361.

Verma, R. S., and Dosik, H. (1981a). *Experientia* **37**, 241.

Verma, R. S., and Dosik, H. (1981b). *Hum. Genet.* **56**, 329.

Verma, R. S., and Lubs, H. A. (1975). *Hum. Genet.* **30**, 225.

Verma, R. S., and Lubs, H. A. (1976a). *Hum. Hered.* **26**, 315.

Verma, R. S., and Lubs, H. A. (1976b). *Can. J. Genet. Cytol.* **18**, 45.

Verma, R. S., Dosik, H., and Lubs, H. A. (1977a). *J. Hered.* **68**, 262.

Verma, R. S., Dosik, H., and Lubs, H. A. (1977b). *Hum. Genet.* **38**, 231.

Verma, R. S., Dosik, H., and Lubs, H. A. (1977c). *Ann. Hum. Genet.* **41**, 257.

Verma, R. S., Benjamin, C., Rodriguez, J., and Dosik, H. (1981). *Hum. Genet.* **59**, 412.

Verma, R. S., Shah, J. V., and Dosik, H. (1982). *Cytobios* **34**, 175.

Verma, R. S., Benjamin, C., and Dosik, H. (1983). *Cytobios* **37**, 157.

Verma, R. S., Shah, J. V., and Dosik, H. (1984). *Stain. Tech.* **59**, 13.

Verma, R. S., Babu, K. A., Chemitiganti, S., and Dosik, H. (1985). Submitted.

Wachtler, F., and Musil, F. (1979). *Stain. Tech.* **54**, 265.

Wachtler, F., Ellinger, A., and Schwarzacher, H. G. (1980). *Cell Tissue Res.* **213**, 351.

Wachtler, F., Schwarzacher, H. G., and Ellinger, A. (1982). *Cell Tissue Res.* **225**, 155.

Warburton, D., and Henderson, A. S. (1979). *Cytogenet. Cell Genet.* **24**, 168.

Warburton, D., Atwood, R. C., and Henderson, A. S. (1976). *Cytogenet. Cell Genet.* **17**, 221.

Williams, W. A., Kleinschmidt, J. A., Krohne, G., and Franke, W. W. (1982). *Exp. Cell Res.* **137**, 341.

Wimber, D. E., and Steffensen, D. M. (1973). *Science* **170**, 639.

Wolgemuth-Jarashow, D. J., Jagiello, G. M., and Henderson, A. S. (1977). *Hum. Genet.* **36**, 63.

Wolgemuth, D. J., Jagiello, G. M., and Henderson, A. S. (1979). *Exp. Cell Res.* **118**, 181.

Woodruff, K. M., and Deleon, P. A. M. (1982). *Hum. Genet.* **61**, 27.

Young, B. D., Hell, A., and Birnie, G. D. (1976). *Biochim. Biophys. Acta* **454**, 539.

Zakharov, A. F., Davudov, A., Egolina, N. A., and Benjusch, V. A. (1980). *In* "The European Society of Human Genetics" (Human cytogenetics papers), p. 34. Dubrovnik Symp.

Zakharov, A. F., Davudov, A. Z., Benjush, V. A., and Egolina, N. A. (1982). *Hum. Genet.* **60,** 24.
Zang, K. D., and Back, E. (1969). *Cytogenetics* **37,** 304.
Zankl, H., and Bernhard, T. S. (1977). *Hum. Genet.* **37,** 79.
Zankl, H., and Hahmann, S. (1978). *Hum. Genet.* **43,** 275.
Zankl, H., and Zang, K. D. (1974). *Humangenetik* **23,** 259.
Zankl, H., and Zang, K. D. (1979). *Hum. Genet.* **52,** 119.
Zankl, H., Mayer, C., and Zang, K. D. (1980). *Hum. Genet.* **54,** 111.
Zankl, H., Huwer, H., and Zang, K. D. (1982). *Cancer Genet. Cytogenet.* **6,** 47.

INTERNATIONAL REVIEW OF CYTOLOGY, VOL. 94

Sertoli Cell Junctions: Morphological and Functional Correlates

Lonnie D. Russell and R. N. Peterson

Department of Physiology, School of Medicine, Southern Illinois University, Carbondale, Illinois

I. Introduction

Perhaps no other mammalian cell is as rich in the diversity and numbers of junctions it develops as is the Sertoli cell. The cyclic nature of spermatogenesis and continually changing morphology of the Sertoli cell provide a dynamic framework in which junctions are periodically formed and eliminated. Unlike events in many other organ systems, the numerous changes taking place in the testis during the spermatogenic process are capable of being precisely timed (e.g., Clermont *et al.*, 1959) and the relationship of Sertoli cells to other cells in cell associations may be determined. These features allow for a more precise correlation of structure with function or, at least, anticipated function.

In this review junctions are defined more liberally than their classical descriptions for other epithelia (Farquhar and Palade, 1963) and several junctions are described in which experimental data are the primary evidence to suggest a

177

junctional relationship between cells. We have relied on the text of this review to describe junctions and have avoided repetitiously adding figures which have been published elsewhere. Instead, we have employed the strategy of parenthetically referring to key published figures from the work of others, and selecting for use herein only those figures which portray heretofore unpublished observations or serve to emphasize areas which are in need of additional emphasis.

Information relating to Sertoli cell junctions in mammals is more complete and the bulk of this review is directed toward this class. In nonmammalian species the somatic cell–Sertoli cell population differs dramatically from mammals and wide variations occur among nonmammalian groups. Such a topic well deserves a review of its own.

II. Sertoli Cell–Connective Tissue Junctions

The basal aspect of the Sertoli cell is defined as resting on the basal lamina (Wong and Russell, 1983; Weber et al., 1983). With the exception of spermatogonia and very young spermatocytes, the Sertoli cells are the only cells which reside on the inner surface of the basal lamina. In this area of contact they form rudimentary *hemidesmosomes* with the acellular basal lamina. Reported by Connell (1974) and demonstrated by Russell (1977b; his Fig. 5), hemidesmosomes were seen only at the interface of Sertoli cells with the basal lamina and not between germ cells and the basal lamina.

The hemidesmosome is widely recognized in various tissues as the site of cell and connective tissue adhesion (Kelly, 1966). The scanty attention given this junctional type in the testis would lead one to believe it is of only minor functional importance. On the contrary, we suggest that this junction is the single most important component in maintaining the integrity of the seminiferous epithelium. Sertoli cells are irregularly columnar in shape (Wong and Russell, 1983) and extend to the lumen of the seminiferous tubule, which in most mammalian species is a distance of 60–90 μm (Weber et al., 1983). Adhesion of this columnar cell to the basal lamina by hemidesmosomes provides an affixed Sertoli scaffolding, extending into the seminiferous tubule, to which germ cells are attached, compartmentalized, or buried within deep crypts of that cell. Such an inconspicuous junction is apparently very strong. Hypertonic fluids cannot separate Sertoli cells from the basal lamina, through a mechanism involving cell shrinkage, although germ cells residing on the basal lamina are sometimes pulled away from it by this treatment (Fig. 18 of Gilula et al., 1976). The Sertoli cells must also resist the stresses of tubular contractions produced by myoid cells of the testicular capsule (Leeson, 1975) and the myoid cells of the wall of the seminiferous tubule (Clermont, 1958). Sertoli cells detached from the basal lamina in the normal spermatogenic process have not been described, although this feature may occur in pathological states (Figs. 15–17 of Russell and Gardner, 1974).

III. Sertoli Cell–Sertoli Cell Junctions

A. Tight or Occluding Junctions

The initial report describing tight junctions between Sertoli cells was that of Nicander's (1967), although the figure utilized to illustrate them (Fig. 8 of Nicander) is most probably a gap junction. Others of Nicander's figures (Figs. 2–4) would have been appropriate. In 1970, Dym and Fawcett clearly depicted this junctional type in thin sections, noting its location near the base of the cell at its lateral surface. These authors described focal sites of apparent membrane fusion which restricted the passage of intravascularly injected lanthanum nitrate toward the lumen of the seminiferous tubule. Several other electron-dense substances of varying molecular weights were shown to be retarded by occluding junctions (Fawcett et al., 1970). Tight junctions were also shown to restrict the passage of hyperosmotic fluids toward the tubular lumen and these fluids prevented cell shrinkage distal to the focal tight junctions (Gilula et al., 1976; Russell, 1977b).

Ross (1970) demonstrated in the same region where Sertoli tight junctions were present that adjacent Sertoli cells form a continuous contact zone around the seminiferous tubule which could potentially segregate one portion of the seminiferous tubule from another. From the thin section studies of this region Dym and Fawcett (1970) found that they could not determine whether junctions were punctate or represented linear bands of contact that extended for some distance. This question was answered simultaneously in two separate laboratories (Gilula et al., 1976; Nagano and Suzuki, 1976b) where freeze-fracture techniques were employed to view split membrane surfaces and the junctional particles forming Sertoli–Sertoli occluding junctions. Occluding junctions appeared as rows of intramembranous particles preferentially associated with the E face of the membrane and not with the P face, as is usually the situation in other epithelia. The P face demonstrated up to 40 (Gilula et al., 1976) or 60 (Nagano and Suzuki, 1976b) tight junctional rows, the majority of which were nonanastomotic. In our experience, expansive fractures of rat Sertoli cells reveal over 50 rows of occluding junctions and their number may be over 100 (Fig. 1). The linearity exhibited by tight junctions may also be demonstrated using osmium:ferrocyanide fixed testis (Russell and Burguet, 1977), since under these fixation conditions junctional particles appear clearly defined as electron-translucent spots within each

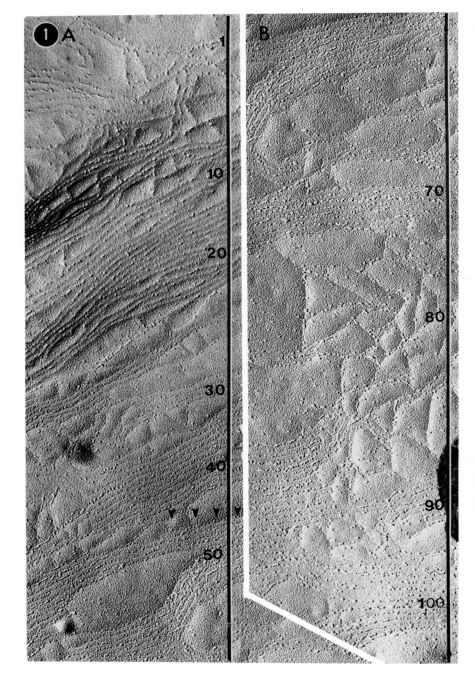

bilayer half of the membrane (Fig. 2). *En face* sections of Sertoli occluding junctions outlined with electron-dense tracers also reveal to advantage the linearity of junctions (Figs. 8–10 of Neaves, 1973).

Since the initial observations on Sertoli–Sertoli junctions in the rat, various investigators have confirmed their presence in other species (Dym, 1973; Nagano and Suzuki, 1976a,b; Connell, 1978; Gondos and Connell, 1978; Osman and Plöen, 1978; Nagano, 1980; Pelletier and Friend, 1983). Nagano *et al.* (1982), using freeze-fracture, have reported that in the mouse the Sertoli occluding junctions are atypical from those in other epithelia in that they remain unaffected by fixation (i.e., glutaraldehyde-fixed and fresh-frozen tissues appear identical with regard to particle distribution). Recently, Pelletier and Friend (1983) have demonstrated, using both freeze-fracture and thin section methodologies, regional differences in the junctional properties at the basolateral aspect of the guinea pig Sertoli cell. The most basally positioned junctions were uninterrupted parallel strands referred to as "continuous zonules." Further toward the lumen the junctional strands were meandering and incomplete and referred to as "discontinuous zonules."

The Sertoli–Sertoli cell occluding junctions, like occluding junctions in other tissues, are believed to function in maintaining a permeability barrier. The permeability barrier at Sertoli occluding junctions is thought, by some, to be especially tight in view of the large numbers of parallel rows of junctions circumscribing each cell, although Martinez-Palomo and Erlij (1975) suggested that the numbers of rows of occluding junctions are not necessarily related to permeability. The occluding junctions separate two populations of cells. Near the periphery of the tubule, spermatogonia and young spermatocytes are said to reside in a *basal compartment,* whereas mature germinal cells are situated on the other side of the tight junctions and are said to reside in the *adluminal compartment* (Dym and Fawcett, 1970; Russell, 1977b).

The ability of occluding junctions to exclude dyes from penetrating to the lumen of the seminiferous tubule has been known for many years (Ribbert, 1904; Bouffard, 1906; Pari, 1910). Detailed studies in Setchell's lab have demonstrated the exclusion or retardation of physiological substances from the seminiferous tubule fluid (Johnson and Setchell, 1968; Setchell *et al.,* 1969; Setchell, 1974). Only recently have the occluding junctions been shown to exclude material introduced into the tubular lumen (Ross, 1977; Pelletier and Friend, 1983).

FIG. 1. Freeze fracture of the Sertoli–Sertoli occluding junctions demonstrating rows of intramembranous particles associated with adjacent Sertoli cells. Some junctional strands are parallel and circumferentially arranged whereas others are anastomotic. A and B are continuations of the same figure and the common region in A with B lies below the arrowheads. Over 100 rows of occluding junctions are enumerated along a solid line passing roughly perpendicular to junctional strands. ×27,000.

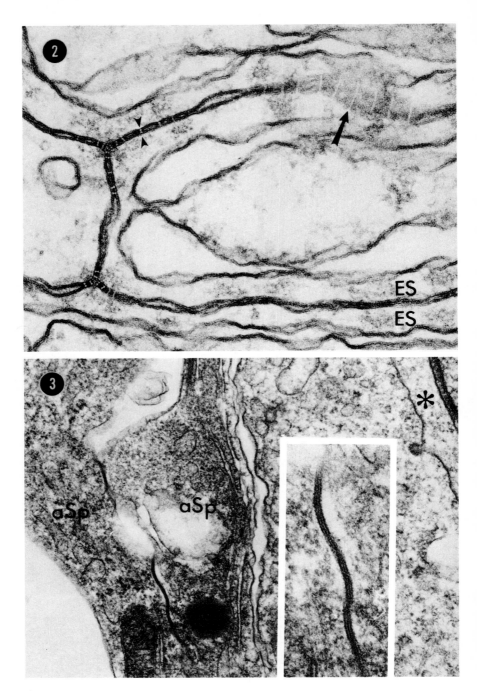

That most vascularly perfused substances of high molecular weight do not reach the tubular lumen or do not readily attain concentration equilibration with seminiferous tubular fluid has given rise to the popular term describing this permeability barrier—the *blood–testis barrier* (Chiquoine, 1964). Willson *et al.* (1973) have pointed out the inadequacy of this terminology since the permeability barrier most commonly resides between the lymphatics and fluid bathing the periphery of the seminiferous tubule and the fluid of the adluminal compartment and/or tubular lumen. Setchell and Waites (1975) concluded that there is comparatively unrestricted passage of substances from the blood capillaries to the lymphatic system. Leydig cells (Lalli and Clermont, 1975) and potentially the seminiferous tubules themselves secrete substances into the lymphatics and the fluid surrounding the seminiferous tubule giving these fluids a unique character in terms of the concentrations of specific substances. Willson *et al.* (1973) have suggested the term *testicular–tubular barrier* in an attempt to correct the inadequacies of the term "blood–testis" barrier. We will employ the term *Sertoli cell barrier* throughout this review. The term implies no prejudice in describing the direction which the barrier maintains a gradient for particular substances. The following options are available to either fluids of the tubular lumen or fluids bathing the tubule: (1) they may be restricted from entering or pass through the series of occluding junctions; (2) they may be excluded by the Sertoli cell; (3) they may enter the Sertoli cell to be actively or passively discharged in a native or modified form to the same compartment; and (4) they may enter the Sertoli cell to be actively or passively discharged in a native or modified form to another compartment. It is obvious that only two components, (1) the Sertoli occluding junctions and (2) the body of the Sertoli cell, are operative in forming the Sertoli cell barrier, thus illustrating the appropriateness of referring to the barrier as a function of the Sertoli cell. There is some perception that the Sertoli cell barrier is very different from barriers elsewhere in the body even though there are not designated names for most of these barriers. The Sertoli cell barrier is not unique or functionally different than epithelial barriers which selectively exclude bidirectional passage of molecules. Use of the term, Sertoli cell

---

FIG. 2. This region of interface of several Sertoli cells postfixed with osmium:ferrocyanide shows occluding junctions as regions of electron translucency within the membrane bilayers. Paired, focal translucencies (apposing arrowheads) are seen when membranes are sectioned perpendicularly and single linear translucencies (isolated arrow) are apparent in *en face* sections. The Sertoli ectoplasmic specialization (ES) flanks the region of Sertoli–Sertoli occluding junctions. ×67,000.

FIG. 3. An apparent gap junction formed between rat Sertoli cells is shown. This junction was found near the lumen of the seminiferous tubule at mid-cycle. The apical Sertoli processes depicted (aSp) are flanked by round spermatids, one of which demonstrates an acrosome (asterisk). ×28,000. The gap junction is shown at higher magnification in the inset. ×64,000.

barrier, directly pinpoints the ''epithelia'' involved and goes a long way toward correcting the implications of the term, blood–testis barrier.

Transfer of materials across the occluding junctions would seemingly be largely dependent on molecular size and physical properties of transferred substances. Movement of molecules across the body of the Sertoli cell is a potentially more selective process. Here, receptors, pumps, exchange mechanisms, enzyme systems, transport, and internalization systems would seem to be operative. The concentration ratios of ions, amino acids, sugars, macromolecules, etc. in the tubular fluid may be greater than or less than in testicular lymph (Setchell and Waites, 1975; Waites, 1977). The gradient, however, for most substances is such that they remain in higher concentrations in testicular lymph than in seminiferous tubule fluid. There is little information regarding transfer of substances from the tubular lumen across the Sertoli cell although we suspect such movement does occur.

Fawcett (1975) has pointed out that the Sertoli cell maintains a fluid gradient with water being moved from the outside of the tubule to the tubular lumen. This polarity, which is reversed from that of other epithelia, is also reflected in the position of the tight junctions which are located near the basolateral aspect of the cell. Most other epithelia demonstrate junctions at their apicolateral surface. The placement of Sertoli tight junctions allows more advanced germinal cells to be bathed in the fluid environment provided by the Sertoli cells. As fluid secreted by the Sertoli cells into the lumen of the seminiferous tubules enters the epididymis, it is reabsorbed by the epididymal epithelium. The position of epididymal epithelial cell occluding junctions, at or near their apicolateral border, reflects a change in the direction of fluid flow. Not only are Sertoli–Sertoli junctions displaced relative to most other epithelia, but the sequence of occluding junctions and adhering junctions (see Section III,D) is reversed (Pelletier and Friend, 1983).

1. Development

Development of occluding junctions has been more extensively investigated in the rat (Vitale *et al.*, 1973; Gilula *et al.*, 1976; Meyer *et al.*, 1977) than in any other single species (dog, Connell, 1980; mouse, Nagano and Suzuki, 1976b; guinea pig, Pelletier and Friend, 1983; and human, Furuya *et al.*, 1978). In embryonic development, disorganized rows of intramembranous particles are seen preferentially on the E face of the Sertoli plasma membrane. Scattered, blind-ending junctional strands are noted in prepubertal animals, but these strands do not extend around the cell (Hadziselimovic, 1977; Gondos, 1980; Camatini *et al.*, 1982; Pelletier and Friend, 1983). At this time, lanthanum freely permeates between all cells of the seminiferous epithelium (Connell, 1980; Furuya *et al.*, 1980a). As the interior of the seminiferous epithelium becomes impermeable to either lanthanum, hypertonic fluids, or other tracers, the junctional rows demonstrate a parallel alignment although many of them continue to anastomose or end blindly. They also become circumferentially arranged around the cell.

A lumen forms within seminiferous tubules coincident with the organization of Sertoli tight junctions in parallel array (Vitale *et al.*, 1973). Pelletier and Friend (1983) have emphasized that the parallel orientation of junctions is only a reflection of the completion of junctional strands and that the latter (formation of "continuous zonules") ensures a permeability barrier. The "discontinuous zonules," which lie near the tubular lumen, remain permeable to tracer (filipin) if the tracer is allowed to permeate from the lumenal aspect of the tubule. The development of tight junctions (presumedly "continuous zonules"; Pelletier and Friend, 1983) also occurs coincident with the initiation of spermatogenesis. Tubules containing cells in the early prophase of meiosis inevitably develop Sertoli–Sertoli occluding junctions (Gondos and Connell, 1978; Cavicchia and Burgos, 1983). It is interesting to compare the timing of occluding junction formation during pubertal development with the timing of occluding junction formation in adult animals as a function of the cyclic maturation of germ cells (see below). In both situations the germinal cells become isolated at precisely the same phase of development. They enter separate compartments during early prophase (leptotene or zygotene) of meiosis (Dym and Cavicchia, 1977; Russell, 1977b, 1978a). Furuya *et al.* (1978), however, reported lanthanum, impermeable, tight junctions in the testis of a boy of 11 years at which time spermatocytes were not present. Pelletier and Friend (1983) showed in the guinea pig that the manner in which "continuous zonules" formed from "discontinuous zonules" was similar in pubertal animals where spermatogenesis is initiated and in adult animals during the cyclic maturation of germ cells.

2. *Cyclic Changes*

In the adult, spermatogenesis proceeds cyclically, the least mature of the germinal cells dividing and differentiating and at a specific time moving away from the basal lamina in a direction toward the tubular lumen. Dym and Fawcett (1970) recognized that it was necessary for maturing germ cells to cross the Sertoli cell occluding junction without disrupting them. Russell (1977b, 1978b) demonstrated in the rat, using hypertonic fixatives and lanthanum tracers, that germ cells cross a permeability barrier just after meiosis is initiated. Similar results were obtained for the monkey (Dym and Cavicchia, 1977). Thus, spermatogonia and very young germ cells reside in the basal compartment. Adjacent Sertoli cells appear active in the coordinated translocation of spermatocytes by insinuating basal processes between the young spermatocytes and the basal lamina. When the basal processes from several adjoining Sertoli cells meet, these processes adhere to each other and shortly thereafter occlude lanthanum. The occluding junctions above spermatocytes remain intact as the new ones form below the spermatocytes. How long they remain intact is not known, but while the young clones of spermatocytes are sequestered in a compartment of their own, they can be said to reside neither in the basal nor in the adluminal compartment, but one which is intermediate between them. The short-lived *intermediate compartment* probably functions as a transit chamber to assure that there is no breach of the Sertoli occluding junctions along the length of the seminiferous tubule (Russell, 1977b, 1978b). Freeze-fracture studies, performed at precisely the period of upward movement of spermatocytes, are necessary to determine the intramembranous changes which accompany the upward movement of spermatocytes.

3. *Development in Vitro*

In cultures of prepubertal rat seminiferous tubules, Sertoli cells develop tight junctions at the time they would have normally done so had they developed *in vivo* (Eddy and Kahri, 1976). Forty days of organ culture of seminiferous tubules from 1-day-old rats did not prevent formation of tight junctions (Meyer *et al.*, 1978). However, it did alter the parallel arrangement of junctional strands. The authors concluded that gonadotropin was not necessary for junction development. The loss of parallel junctional configurations seen *in vitro* (Biglardi and Vegni-Talluri, 1976; Meyer *et al.*, 1978) allows lanthanum to permeate between cells (Biglardi and Vegni-Talluri, 1976). Dissociated Sertoli cells from juvenile animals maintained in culture also develop tight junctions (Tung *et al.*, 1975; Eddy and Kahri, 1976; Cameron and Markwald, 1981; Janecki *et al.*, 1981; Espevic *et al.*, 1982).

4. *Pathology*

Numerous reports have appeared aimed at determining the status of the Sertoli cell barrier and the integrity of Sertoli occluding junctions after experimental treatments or in animals with genetic disorders (Table I). By and large these studies have employed lanthanum or dye penetration, freeze-fracture, or hypertonic fluids to demonstrate the integrity, or lack of integrity, of Sertoli occluding junctions. These tests are crude indicators of transepithelial permeability and demonstrate only gross changes occurring in Sertoli occluding junctions and not the net effect on Sertoli cell permeability, itself. Fluid comparisons appear to offer a more sensitive approach to studying pathological changes in the Sertoli cell barrier (Okamura *et al.*, 1975; Koskimies, 1973; Main and Waites, 1977, etc.) Here, the concentration of a substance in rete testis fluid, or preferably seminiferous tubule fluid, is expressed as a ratio of their blood or lymph concentration. A battery of substances with varying molecular weights and physical properties might be used, in the future, as standards tests where control and experimentals can be compared.

The conclusions reached, as presented in Table I, generally provide negative results indicating that primarily the Sertoli–Sertoli occluding junctions and in some cases the Sertoli cell itself, are reasonably resistant to penetration of tracer under a variety of deleterious conditions. Even in many situations where spermatogenesis was severely disrupted the treatment did not alter the Sertoli occluding junctions. Some studies yielding positive results must be questioned on methodological grounds. The results described after efferent ductule ligation are contradictory; one study indicating damage to the occluding junctions (Neaves, 1973), and others indicating that these junctions remain intact (Ross, 1977; Osman and Plöen, 1978). Of particular interest are those studies relating to the control factors influencing the development of the occluding junctions in pubertal animals. Formation of the occluding junction component of the barrier is delayed (Vitale *et al.*, 1973) or prevented (Bressler, 1976) in the absence of hormones, but its maintenance appears independent of hormonal control (Hagenäs *et al.*, 1977, 1978; Furuya *et al.*, 1980a; Aumüller *et al.*, 1978). The results are at variance with one other study which examined the formation of Sertoli cell junctions in tissue culture media lacking hormones (Eddy and Kahri, 1976). The functional integrity of Sertoli junctions *in vitro* is more difficult to ascertain than their integrity in the intact animal.

B. GAP JUNCTIONS

The first report of Sertoli–Sertoli gap junctions was provided by Nicander (1967). He did not recognize this junctional type as a gap junction for he termed it a "five-layered tight junction or narrow junction." Adjacent Sertoli cells develop gap junctions during fetal life (Nagano and Suzuki, 1978), during prepubertal life (Camatini *et al.*, 1982) and during puberty as spermatogenesis is initiated (Gilula *et al.*, 1976; Nagano and Suzuki, 1976b; Meyer *et al.*, 1977; Camatini *et al.*, 1982). As tight junctions form, gap junctions are found between tight junctional strands. They appear as hexagonally packed particles on the E face of the membrane and as corresponding pits on the P face. In the adult, gap junctions are rarely found in comparison with the developing animal. They are positioned between tight junctional strands (aisle configuration) at the level of the Sertoli–Sertoli occluding junctions (Gilula *et al.*, 1976; McGinley *et al.*, 1977; Camatini *et al.*, 1979, 1982; Connell, 1980). Gap junctions in the adult are also detectable in thin sections (Nicander, 1967; Dym and Fawcett, 1970; Fawcett, 1970; Gilula *et al.*, 1976; Pelletier and Friend, 1983) and with freeze fracture (Gilula *et al.*, 1976; Nagano and Suzuki, 1976b; Meyer *et al.*, 1977; Pelletier and Friend, 1983), although in one study of the normal, adult, human they were not detected (Nagano and Suzuki, 1976a).

With transmission electron microscopy we have also found Sertoli–Sertoli membrane associations with an approximate 2–4 nm intercellular space near the tubular lumen of rat seminiferous tubules (Fig. 3 and inset). These are rarely found and we have only observed them at stages near mid-cycle.

There are few studies on gap junctions in pathological conditions, but those

TABLE I

Integrity of Sertoli Occluding Junctions and/or Sertoli Cell Barrier in Pathological Conditions or Accompanying Genetic Disorders

Investigator and year[a]	Animal	Method utilized	Condition	Finding
Johnson (1970a)	Guinea pig	Dye penetration	Isoimmunization w/testis tubules	Dye penetration in severely damaged tubules
Johnson (1970b)	Rat	Dye penetration	Hypophysectomy	No change
Koskimies (1973)	Rat	Fluid comparisons	Cadmium	Increased protein in reproduction tract fluids
Nagano and Okumura (1973)	Guinea pig	Immune localization	Allergic aspermatogenesis	Breach of barrier in 24 hours
Neaves (1973)	Rat	Tracer[b]	Efferent ductule ligation	Junctions permeable to lanthanum
Vitale et al. (1973)	Rat	Tracer	Estrogen, clomiphene	Junction formation delayed
Willson et al. (1973)	Guinea pig	Tracer	Isoimmunization w/testis	Focal breach of occluding junctions
Castro and Seiguer (1974)	Guinea pig	Tracer	Immune aspermatogenesis	No change
Aumuller et al. (1975)	Monkey	Tracer	Cyproterone	No change
Vitale et al. (1973)	Rat	Tracer	Sertoli cell-only syndrome induced by irradiation	Delay in formation of occluding junctions
Bressler (1976)	Mouse	Transmission electron microscopy	Newborn testis transplanted into testis of normal and hypophysectomized hosts	Failure of occluding junctions to develop in hypophysectomized hosts
Gilula et al. (1976)	Rat	Freeze-fracture	Sertoli cell only syndrome induced by Busulfan	No change
Gravis et al. (1976)	Hamster	Tracer	Epinephrine	No change
Hagenäs and Ritzen (1976)	Rat	Tracer	Epinephrine	No change
Nagano and Suzuki (1976a)	Rat	Freeze-fracture	Testicular feminization	No change
Hagenäs et al. (1977b)	Rat	Tracer	Experimental cryptorchidism	No change

Reference	Species	Method[b]	Treatment	Result
Main and Waites (1977)	Rat	Fluid comparisons	Increased temperature	Minor changes in rete testis fluid
Nagano et al. (1977)	Mouse	Freeze-fracture	Germ cell free testis	No change
Ross (1977)	Mouse	Tracer	Efferent ductule ligation	No change
Aumüller et al. (1978)	Rat	Freeze-fracture	Hypophysectomy and testosterone treatment	No change
Hageäs et al. (1978)	Rat	Tracer	Hypophysectomy	No change
Meyer et al. (1978)	Rat	Freeze-fracture	Tissue culture (30–40 days)	Junctions develop without gonadotropins
Neaves (1978)	Rat	Tracer	Vasectomy	Areas of breach of junctions
Osman and Plöen (1978)	Rat	Tracer	Efferent ductule ligation	No change
Cameron and Snydle (1980)	Human	Tracer	Varicocele	No change
Furuya et al. (1980a)	Human	Tracer	Hypogonadotropic eunuchoidism and postpubertal pituitary failure before and after HCG treatment	Occluding junction development affected, but maintenance not affected
Pelletier and Friend (1980)	Guinea pig	Filipin	Gossypol	Disruption of junctions
Camatini et al. (1981)	Human	Tracer	Germinal aplasia	No change
Camatini et al. (1982)	Human	Tracer and freeze-fracture	Lymphoblastic leukemia and chemotherapy	No change
Espevik et al. (1982)	Rat	Freeze-fracture	Cadmium in vitro	No change
Cavicchia and Burgos (1983)	Rat	Tracer	Busulfan	Developmental failure of junctions to become oriented
Franchi and Camatini (1983)	Guinea pig	Thin section and freeze-fracture	Trifluoroperazine	Disruption of occluding junctional strands and ES
Fritz et al. (1983)	Mice	Hypertonic fluids	Testicular immunization	No change

[a]Selected studies since 1970.
[b]"Tracer" refers to a substance which appears electron dense.

available suggest that neither hypophysectomy (Aumüller *et al.*, 1978) nor infertility (Bigliardi and Vegni-Talluri, 1977) alters the number of gap junctional particles between Sertoli cells.

Gap junctions are widely accepted as the morphological entities by which certain forms of cell-to-cell communication (ionic or metabolic coupling) can occur. Electrical coupling between Sertoli cells has been recently demonstrated (Eusebi *et al.*, 1983). The pubertal and prepubertal animal, having extensive areas of gap junctional contact, record low resistance levels during electrical coupling. In the adult, which demonstrates fewer gap junctional areas, the resistance is much greater. Intuitively, the converse would be suspected on the basis that increased cooperation between adult Sertoli cells would be necessary to coordinate the activity of adjacent Sertoli cells during the cyclic events taking place during spermatogenesis.

C. Septate-Like Junctions

Gilula *et al.* (1976) suggested, for the rat, that many of the junctions at the lateral aspect of the Sertoli cell, in the region of the occluding junctions, possessed characteristics of both tight and septate junctions. This suggestion was offered since many of the junctions normally observed, like septate junctions, are seen in pairs and the rows do not freely anastomose. Working with the dog, Connell (1978) argued that paired parallel rows of junctions were true septate junctions and distinctly different from the nearby occluding junctions. Connell found that the septate variety of junctions did not anastomose. They appeared in thin section micrographs as septa crossing the intercellular space and were outlined by, but did not exclude lanthanum, as do tight junctions. Connell suggests that this was the first demonstration of tight junctions in vertebrates. In the dog septate junctions develop prior to tight junctions (Connell, 1980). Camatini *et al.* (1982) demonstrated similar paired parallel rows of septate junctions in the human. Connell (1980) suggests that the septate junctions are present in several species. Nagano and Suzuki (1983) are skeptical that the paired junctional rows are septate and find it difficult to compare the mammalian Sertoli "septate" junction with those found in invertebrates. Septate junctions were not mentioned in a recent comprehensive freeze-fracture study of guinea pig testis Sertoli junctions (Pelletier and Friend, 1983).

D. Desmosomes

The junctional complex of most epithelia includes a macula (desmosome) and/or zonula adherens (Farquhar and Palade, 1963). Desmosomes are also seen between Sertoli cells at the level of the occluding junctions (Bawa, 1963; Nicander, 1967; Connell, 1974, 1978; Altorfer *et al.*, 1974; Pelletier and Friend, 1983). They are rarely found and their presence is not emphasized in the literature, and even recently their existence has been disputed (Nagano and Suzuki, 1983). In the first of the two forms of the junction (Figs. 4 and 5), there is a dense accumulation of 10-nm subsurface filaments; the intercellular space may be slightly irregular or alternatively very regular. Some desmosomes show a faint intermediate line (Fig. 11 of Pelletier and Friend, 1983). In the second form, subsurface densities and 10-nm filaments are periodically spaced along the interface of two Sertoli cells. In the interval between the periodicities, tight junctions are noted where the plasma membranes converge (Fig. 6). Both forms of desmosomes are most frequently seen at the adluminal extent of the rows of tight junctions (Pelletier and Friend, 1983; Fig. 4).

Not all Sertoli cells possess desmosomes, for none was observed in a recently reconstructed Sertoli cell (Wong and Russell, 1983; Russell *et al.*, 1983a). The functional property of desmosomes, as has been postulated for epithelia elsewhere, is cell-to-cell adhesion. The rare occurrence of these junctions, as well as the presence of numerous other junctional forms which might possess adherent properties, raises the question of whether they serve a function at all, or are simply rudimentary structures.

Even more rarely seen than basolaterally positioned desmosomes are those which are occasionally at the lateral surface of the cell near the lumen. These Sertoli–Sertoli desmosomes have only been found in Type B configuration Sertoli cells (Wong and Russell, 1983; Russell and Karpas, 1983) at a time where the elongate spermatid heads are grasped by the apical expansions of Sertoli cells and the spermatid flagella extend into the tubular lumen. The delicate Sertoli cell expansions, which hold the spermatid heads, are tied together by occasional desmosomes (Figs. 13 and 14 of Russell, 1984). These desmosomes are short-lived structures for, during sperm disengagement, the Sertoli cell reverts back to a Type A configuration. It seems possible that the apically positioned desmosomes secure the delicate lumenal extensions of Sertoli cells.

E. ECTOPLASMIC SPECIALIZATIONS (ES)

By itself, this subsurface specialization of the Sertoli cell should not necessarily be classified as a junction, but because it is often associated with junctions of one type or another, it is frequently referred to as a ''junctional specialization'' (Flickinger and Fawcett, 1967). Its presence was described at two sites within the testis; at the level of Sertoli–Sertoli occluding junctions (Burgos and Fawcett, 1955; Brökelmann, 1961, 1963; Flickinger and Fawcett, 1967; Nicander, 1967) and facing germinal elements (see Section IV,D).

ES between Sertoli cells at the basolateral aspect of the Sertoli cell may be envisioned as a specialized subsurface region which, like a band in three dimen-

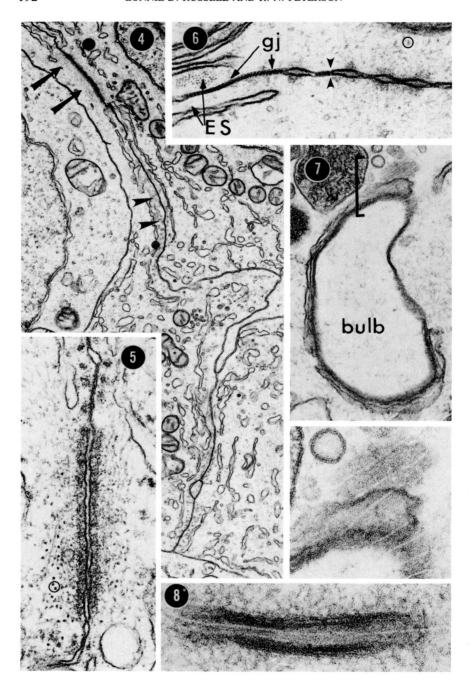

sions, encircles the Sertoli cell (Figs. 1–5 of Russell *et al.*, 1983b; Weber *et al.*, 1983). ES may or may not be present at sites where the aforementioned junctions are seen (except desmosomes). The band is flattened like a sheet or belt and is formed by subsurface bundles of hexagonally packed actin filaments (Toyama, 1976; Franke *et al.*, 1978) and more deeply placed flattened saccules of endoplasmic reticulum showing occasional ribosomes on their deep surface. Adjacent Sertoli cells generally display complimentary positioned sheets of ES. Like ES facing germinal cells (see Section IV,D), that facing adjoining Sertoli cells are hypothesized to be a mantle which is linked to the cytoskeletal components of the body of the Sertoli cell. The actin filaments show links (some of which may be α-actinin) to both the plasma membrane and endoplasmic reticulum (Russell, 1977c; Franke *et al.*, 1978). It is not clear if there is myosin among actin filament bundles (Franke *et al.*, 1978). ES shows close association with microtubules. These lie either deep to the endoplasmic reticulum or among filament bundles (Fig. 10-8 of Fawcett, 1970).

ES is first seen as occluding junctions form and become capable of excluding lanthanum. During the cyclic events of adult spermatogenesis, they apparently form *de novo* at approximately the same time as the junctions form below upward migrating spermatocytes to create an intermediate compartment (see Section III,A,2) and as the Sertoli junctions below upward moving spermatocytes show lanthanum exclusion properties (Russell, 1977b, 1978b). Thus during pubertal development and adult spermatogenesis, ES forms in conjunction with spermatocyte maturation.

It is difficult to assign a function to ES since there is general uncertainty in the literature as to whether ES is a junctional type, is only associated with junctions and imparts certain properties to them, or is simply a subsurface structure seen near junctions, but has little or nothing to do with them. Several investigators believe that ES is fundamentally similar in function to a zonula adherens or desmosome (Brökelmann, 1963; Flickinger and Fawcett, 1967; Nicander, 1967)

Figs. 4–6. Desmosomes formed between adjacent Sertoli cells in the monkey (Figs. 4 and 5) and rat (Fig. 6). The desmosome (arrows) of Figure 4 is found in regions of the interface of Sertoli cells not lined by ES (arrowheads). ×15,000. At higher magnification (Fig. 5; ×68,000) this type of desmosome shows an intercellular space of about 10–15 nm and subsurface densities and more deeply positioned filaments (encircled) which are about 10 nm across. A second type (Fig. 5) of desmosome is characterized by periodic subsurface densities and scattered 10-nm filaments (encircled). Between densities the Sertoli membranes converge to form occluding junctions (paired arrowheads). A gap junction (gj), showing regularly spaced translucencies, is indicated. ES lines the interface of adjacent Sertoli cells; however, the ES does not appear flanking the occluding junctions in the region of the desmosome. ×78,000.

Figs. 7–8. Sertoli–Sertoli tubulobulbar complexes showing *en face* views (see Fig. 2 above) of occluding junctions (arrowheads) in both the bulbous (Fig. 7) and tubular (Fig. 8) portions of the complex. ×38,000, ×126,000, respectively. Below Fig. 7 is an enlargement (×99,000) of the bracketed portion of the figure.

and may be important in holding adjacent Sertoli cells together. After ligation of the ductuli efferentes, the interface between Sertoli cells was not opened by back pressure in the seminiferous tubules (Ross, 1977), suggesting that the ES may regulate cell adhesion. The failure of this area to open might also be due to junctions (tight or gap) present at this site. This hypothesis is intuitively appealing, but awaits further experimental testing. Recently, Franchi and Camatini (1983) have suggested that the ES and occluding junctions are an integrated unit since both are affected by the calmodulin antagonist, trifluoroperazine. The possible relationship between actin filaments at ectoplasmic specialization sites and occluding junctions is not known.

A second, but not wholly unrelated or uncomplimentary hypothesis was brought forward by this author (Russell, 1977c), through investigations primarily aimed at looking at Sertoli–germ cell ESs (see Section IV,D below). The ES was considered an integrated unit or mantle which could be acted upon by cytoskeletal structures extending deep within the Sertoli cell. Through this mantle, the position, arrangement, or movement of that surface portion of the Sertoli cell which forms a band around the cell could be regulated by internal mechanisms. There appears to be a need for such regulation. Clones of germ cells connected by intercellular bridges reside on the basal lamina and during maturation are displaced toward the lumen between adjacent Sertoli cells. Such a complex clone of cells must be positioned *between Sertoli cells* and not entirely covered by these cells or the process of upward movement could not occur. Thus the interface of adjoining Sertoli cells must be positioned over the germ cells which are due to be moved upward. A role for the ES in this process might be presumed, but like the first of the theories mentioned above, this hypothesis awaits experimental verification.

F. CLOSE JUNCTIONS

Expanses of Sertoli–Sertoli contact, in which the intercellular space is very uniform and greatly reduced in width (7–10 nm) from unmodified areas of cellular contact, have been described (Brökelmann, 1963; Flickinger and Fawcett, 1967; Dym and Fawcett, 1970; Fawcett, 1970). These expanses of Sertoli–Sertoli contact which occur at the level of Sertoli occluding junctions have been termed "close junctions" (see Fig. 10-8 of Fawcett, 1970) or "narrow junctions" (Dym and Fawcett, 1970) and are invariably found in areas lined by ectoplasmic specialization (ES). Examination of the intercellular space, after the usual glutaraldehyde:osmium fixation reveals no feature to suggest that this region is a junction per se. However, a certain degree of caution must be exercised if one is to exclude the close junction as a potential adhesive device or a device with other functional properties. First, Connell (1978) finds that septate-like junctions are present in regions where the intercellular space is reduced to 9 nm. Special morphological techniques may be necessary to visualize such junctions. Second, the reader is referred to Section IV,D of this review where Sertoli ES facing the head region of elongate spermatids is described. Here, in spite of a relatively unmodified and reduced (<15 nm) intercellular space, the two cells appear firmly linked together and trypsin treatment is necessary to separate them (Sapsford, 1963; Romrell and Ross, 1979). The similarities in properties of close junctions and the junctions facing elongate spermatids should be considered in future attempts to ascertain a role for "close junctions."

G. Sertoli–Sertoli Tubulobulbar Complexes

In the general region of Sertoli–Sertoli junctions, an elaborate configurational relationship, apparently unique to the testis, is found in several species (Figs. 7–14) including humans (Figs. 15–16). This relationship is commonly seen in the rat (Figs. 7–8) and has been most extensively described in this species. Termed the *tubulobulbar complex* (Russell and Clermont, 1976; Russell, 1979c), a structure named similarly and with similar appearances has also been found occuring between Sertoli cells and late spermatids (see Section IV,E). As the name implies, the tubulobulbar complex takes the form of a long, narrow tube (2–4 μm) with a terminal bulbous component. Adjacent Sertoli cells participating in forming this structure are separated by a 4–5 nm space in both the tubular and bulbous components of the complex. The tubular portion of the complex is surrounded by fine filaments and the bulbous component often displays a flattened saccule of endoplasmic reticulum abutted against the bulb. Both the tubular and bulbous portions of the complex are regionally modified and display tight (Figs. 7 and 8) and gap junctions.

In the rat, tubulobulbar complexes first form during the initiation of spermatogenic activity at the time occluding junctions first display the capacity to restrict lanthanum. The first sign of a forming tubulobulbar complex is the appearance of a bristle-coated pit with a small process from an adjoining Sertoli cell protruding into this pit (Fig. 1 of Russell, 1979c). There is evidence of fine fibrils extending from the tip of the evaginating process into the coated pit. Such activity is reminiscent of the well-known process of receptor mediated endocytosis, with the unique feature being the uptake of a portion of a cell rather than individual molecules. Subsequently, the tubular portion of the complex elongates, dragging the evagination process with it, and develops a terminal bulbous dilation. The bristle-coated pit is frequently not seen in mature complexes and its fate is unknown.

In the postpubertal rat there is evidence that the numbers of tubulobulbar complexes vary during the spermatogenic cycle, reaching a peak just prior to mid-cycle (stage IV or VI) and soon (stage VII–VIII) decreasing to about one-fifth their former number. They remain at this low number for the remainder of

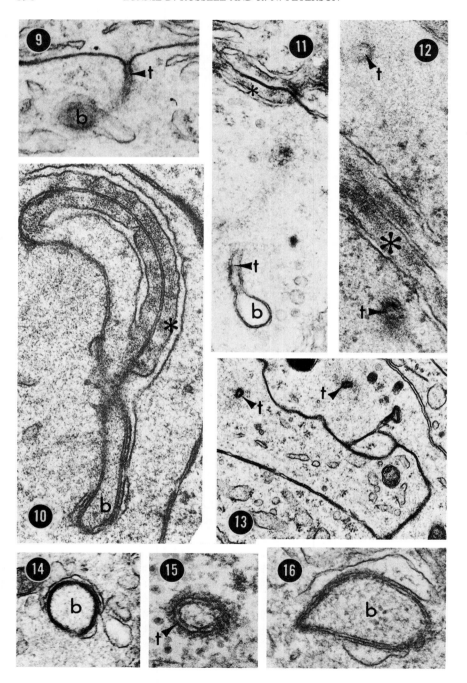

the cycle and early into the next cycle. Morphological and cytochemical evidence was provided showing that most tubulobulbar complexes are degraded at mid-cycle through the activity of Sertoli lysosomes.

The function of tubulobulbar complexes and the implications of the cyclic degradation and formation of complexes is not known. This reviewer (Russell, 1979c) has suggested that tubulobulbar complexes, to a certain degree resemble annular gap junctions (see Larsen, 1977) which are internalized gap junctions, i.e., one cell has internalized a portion of a second cell's cytoplasm and plasma membrane at a gap junction site. In the case of tubulobulbar complexes, both tight and gap junctional membranes are internalized by a single Sertoli cell. As there are no annular gap junctions within the seminiferous epithelium, the tubulobulbar complex may be the comparable morphological device which degrades old junctional contacts. This activity, occurring near mid-cycle, precedes the upward movement of spermatocytes and at least partial junctional breakdown may be a necessary prerequisite for such movement.

IV. Sertoli–Germ Cell Junctions

In spite of early morphological work which indicated that junctions were present between Sertoli cells and germ cells (Nagano, 1959; Brökelmann, 1963; Nicander, 1967), there was a prevailing notion that Sertoli–germ cell junctions were nonexistent. The rationale for this view was conceived through reasoning that germ cells must be free of junctions or enduring junctions to accomplish their movement from the base of the seminiferous epithelium to their eventual release at the tubular lumen (Fawcett, 1970, 1974, 1975). Gradually, this idea was dispelled by the abundance of evidence which demonstrated the presence and, in some cases, functional properties of Sertoli–germ cell junctions. The data suggest that certain of these junctions appear to facilitate, rather than retard, the upward movement of germ cells.

A. DESMOSOMES

Epithelial cells in general, and especially those of the epidermis, demonstrate well-developed desmosomes which have been characterized primarily from this particular epithelium (Kelly, 1966). Less elaborate forms of desmosomes exist in

---

FIGS. 9–16. Sertoli–Sertoli tubulobulbar complexes in various species. Identified in these figures are the tubular portion (t) and bulbous portion (b) of complexes as well as ES (asterisk) which is characteristically found at the interface of Sertoli cells in this general region. Fig. 9, cat, ×24,000; Fig. 10, dog, ×54,000; Fig. 11, mouse, ×45,000; Fig. 12, gerbil, ×57,000; Fig. 13, guinea pig, ×25,000; Fig. 14, monkey, ×43,000; Figs. 15 and 16, human, ×99,000 and ×81,000.

a variety of other tissues and these serve essentially the same function of cell-to-cell adhesion. In the seminiferous epithelium, desmosomes are formed between Sertoli cells and spermatogonia, spermatocytes, and spermatids. First described by Nagano (1959) in fowl, subsequent descriptions for various mammalian species, including man, were provided by others (Brökelmann, 1963; Nicander, 1967; Altorfer et al., 1974; Kaya and Harrison, 1976; Russell, 1977a; Gravis, 1979; Pelletier and Friend, 1983). Desmosomes between Sertoli cells and germ cells are asymmetric structures in that fewer 10-nm filaments (tonofilaments) are associated with the germ cell component at the junctional site than are seen on the Sertoli counterpart of the junction (Figs. 2 and 3 of Russell, 1977a). The junction generally shows a weak intermediate line.

Functionally, desmosomes in the testis are like desmosomes in other tissues acting to hold cells together and maintaining the integrity of the epithelium. This property was demonstrated utilizing hypertonic fixatives which shrink cells and thus cause them to be pulled apart. Under such conditions, cells pull apart in all regions except where desmosomes are present (Kaya and Harrison, 1976; Russell, 1977a,b). Extremely strong hypertonic solutions shrink cells so forcefully that germ cells frequently tear their membranes at desmosome sites as they shrink away (Figs. 9 and 10 of Russell, 1977a). Collagenase and trypsin treatment is capable of dissociating some desmosomes (Russell and Ross, 1979).

Data regarding development, size, numbers, and distribution of desmosomes in the seminiferous epithelium suggest that desmosomes are dynamic structures. For example, in the rat and several other species, desmosomes between Type A spermatogonia and Sertoli cells are relatively rare. As Type A spermatogonia divide to form Type B spermatogonia and these in turn divide to form prelep-totene spermatocytes, the desmosomes are more numerous, expansive, and more highly developed. Leptotene, zygotene, and young pachytene spermatocytes show numerous, small desmosomes connecting adjacent Sertoli cells. The desmosome is most expansive and highly developed in its relationship to mature pachytene and diplotene spermatocytes. Round spermatids and Sertoli cells display small desmosomes and these are only rarely seen. At spermatid elongation all desmosomes are lost. The numbers of desmosomes displayed by a single reconstructed Sertoli cell has been recently quantitated (Russell et al., 1983b). One Sertoli cell demonstrated 19 desmosomes. This Sertoli cell and neighboring Sertoli cells were attached to almost every germ cell adjoining the reconstructed cell. Most desmosomes were positioned in a manner that would resist forces tending to dislodge spermatids in a direction toward the tubular lumen. Sertoli–germ cell desmosomes are also capable of developing in vitro (see Figs. 6 and 7 of Ziparo et al., 1980).

The question of germ cell mobility in view of the presence of desmosomes has been addressed (Russell, 1977b). Two hypotheses could explain how germ cells move upward. First, desmosomes may be capable of forming and dissociating rapidly, allowing short periods when germ cell movement might occur. This hypothesis does not appear to fit the morphological observations of continually adherent desmosomes as cells move from the basal to the intermediate compartment of the seminiferous tubule (Russell, 1977b). The second hypothesis implies that desmosomes, although changing their character from time-to-time, are always fundamentally adherent structures and that because of this property, they actually facilitate upward movement of germ cells (Russell, 1977b, 1980b). Through changes in the configuration of the Sertoli cell, the Sertoli cell influences the position of the germ cell while the desmosome maintains the germ cell in register with a particular area of the Sertoli cell. As most germ cell types are positioned at the lateral aspect of two or more adjoining Sertoli cells, we envision adjoining Sertoli cells cooperating in the movement of germ cells. As a specific example of how this might work, we refer to the movement of clones of young leptotene spermatocytes several microns away from the basal lamina to enter the intermediate compartment of the seminiferous tubule (Russell, 1977b). Adjacent Sertoli cells cooperate by simultaneously sending basal processes below leptotene cells to insinuate themselves between the germ cells and the basal lamina. With time these processes thicken and form junctions which exclude lanthanum (Russell, 1978b). During this time the leptotene spermatocyte remains in register with the Sertoli cell membrane via desmosomes. The desmosomes assure that the germ cells are affected by configurational changes in the Sertoli cell.

Desmosomes are also undoubtedly important in maintaining the integrity of the seminiferous epithelium, a function which should not be underemphasized, since many germ cell types appear only loosely arranged in the seminiferous epithelium. The disappearance of desmosomes during elongation of spermatids suggests that another holdfast device such as ectoplasmic specializations becomes operative at that time (see Section IV,D).

B. GAP JUNCTIONS

This junctional type is, at best, poorly represented in the seminiferous epithelium. In thin sections, gap junctions, although not recognized as such, were first demonstrated in association with desmosomes (Russell, 1977a). Gap junction particles were first observed with freeze-fracture as a component of germ cell membranes (McGinley et al., 1977), but the authors stated that they could not be certain whether the junctions formed between germ cells or between germ cells and Sertoli cells. Later, McGinley et al. (1979) provided both freeze-fracture and thin section micrographs conclusively showing this junction to be between germ cells and Sertoli cells. In thin sections this junction is identified by the narrowing of the intercellular space to 2–4 nm. Freeze-fracture reveals up to 55 gap junction particles on the P face and complementary pits on the E face. All germ cell types show gap junctions although they appear most frequently associated with cells such as pachytene spermatocytes where desmosome development is most elaborate. Desmosomes appear to dissociate as spermatids begin their elongation phase of development. Recently, Pelletier and Friend (1983) have confirmed the observations made in the rat by showing Sertoli–germ cell gap junctions to also be present in the guinea pig. In *in vitro* studies, Ziparo *et al.* (1980) demonstrated that cultured sheets of Sertoli cells adhere to a freshly supplied germ cell. Close examination of their Fig. 7 suggests that gap junctions also form between the two dissimilar cell types *in vitro*.

As discussed above (see Section III,B), gap junctions are thought to function under the broad category of cell-to-cell communication. An operative means for communication within the seminiferous epithelium is considered important for coordination of the spermatogenic process and possible transfer of small molecules from the Sertoli cell to the germ cells. However, these remarks based on intuition must be substantiated in the future with specific data which will firmly establish the role of gap junctions. Concrete examples of transfer between cells have been provided which show that [^{3}H]choline is transferred from Sertoli cells to germ cells *in vitro* (Ziparo *et al.*, 1981) and that Sertoli cells are ionically coupled (Eusebi *et al.*, 1983). Once having been transferred to a particular germ cell, substances may be transferred via intercellular bridges to germ cells of the same clone.

C. Tight Junctions

Russell (1977a), using the rat, demonstrated in regions of Sertoli–germ cell desmosomes that the intercellular space was reduced greatly. In some micrographs (Fig. 14 of McGinley *et al.*, 1979), the intercellular space was found to be obliterated. Freeze-fracture of these regions showed rows of intramembranous particles (Figs. 11 and 12 of McGinley *et al.*, 1979). In thin section and freeze fracture images of guinea pig Sertoli–germ cell junctions, Pelletier and Friend (1983) clearly demonstrated similar "discontinuous zonules." They were formed between all germ cell types and Sertoli cells and were penetrated by tracers. Friend and Pelletier (1983) consider the basally located Sertoli–germ cell junctions to represent the early phases of formation of Sertoli–Sertoli junctions and those apically positioned Sertoli–germ cell junctions to represent the fragmentation of these junctions. This hypothesis seems unlikely considering it implies that junctions are transposed sequentially from germ cells to Sertoli cells and back to germ cells.

D. Ectoplasmic Specializations (ES)

Similar in structure to the ES which lines the interface of adjacent Sertoli cells at the level of the occluding junctions (see Section III,E), the Sertoli ES facing germ cells is formed by bundles of filaments which course along the surface of the Sertoli cell (Fig. 17) and a more deeply positioned flattened saccule of endoplasmic reticulum which is sparcely associated with ribosomes on its deep surface. McGinley *et al.* (1981) have described regions of "fusion" between superficial layers of the endoplasmic reticulum and the Sertoli plasma membrane. The subsurface filaments are actin (Toyama, 1976) and these appear to be linked to both the Sertoli plasma membrane and the superficial aspect of the endoplasmic reticulum (Russell, 1977c). Demonstration of α-actinin in this region has also indicated to Franke *et al.* (1978) that filament links may be present. ES is found facing all germ cell types of the adluminal compartment. In the relationship of ES with elongate spermatids, the intercellular space is highly uniform and reduced to less than 15 nm (Ross, 1977; Russell, 1977c). Images of broad expanses in the region of the ES-elongate spermatid relationship generally reveal no modification of the intercellular space. Occasionally, septa and/or a fuzzy substance is noted traversing the intercellular space (Figs. 18 and 19), but profiles of these structures are relatively rare. Lanthanum does infiltrate the intercellular spaces at ES sites, but reveals no special features suggesting junctional properties (Ross, 1977).

Strictly speaking, the ES is not a junction, for there is no direct evidence that the subsurface organelles of the Sertoli cell participate in any surface relationship between cells. The properties of the *cell's surface* at ES sites (depending on the germ cell type that it faces) suggest that the ES sometimes is, and sometimes is not, *associated* with surface junctional regions. In a position facing round germinal elements, ES areas do not demonstrate a uniform or even a reduced intercellular space (Russell, 1977c). Hypertonic fluids are capable of shrinking adjacent cells and separating round germ cells from Sertoli cells at ES sites (Russell, 1977c). Mechanical dissociation (in the presence of collagenase) also separates round germ cells and Sertoli cells at regions where ES is present (Romrell and Ross, 1979). The properties associated with ES facing elongate(ing) germinal elements are very different. Neither hypertonic fluids (Fig. 16 of Russell, 1977c; Fig. 14 of Gravis, 1979) nor mechanical disruption, in the presence of collagenase (Figs. 7 and 8 of Romrell and Ross, 1979), can separate the two cell types. Treatments which cause selected Sertoli cells to swell greatly, thereby diluting the density of the cell cytosol and the cellular organelles, do not change the relationship of Sertoli ES to elongate germ cells (Russell, 1977c). It is necessary to treat cells with trypsin to separate elongate germ cells from Sertoli cells at ES sites (Sapsford, 1963; Fig. 1 of Romrell and Ross, 1979). Thus, it appears that there is a substance linking elongate spermatids and Sertoli cells and it is proteinaceous, as evidenced by its trypsin sensitivity. Both Russell (1977c) and Romrell and Ross (1979) agree that at the initiation of elongation, the Sertoli cell acquires the property of strong adherence to spermatids and this takes place between plasma membranes at ES sites. The mechanism by which adhesion

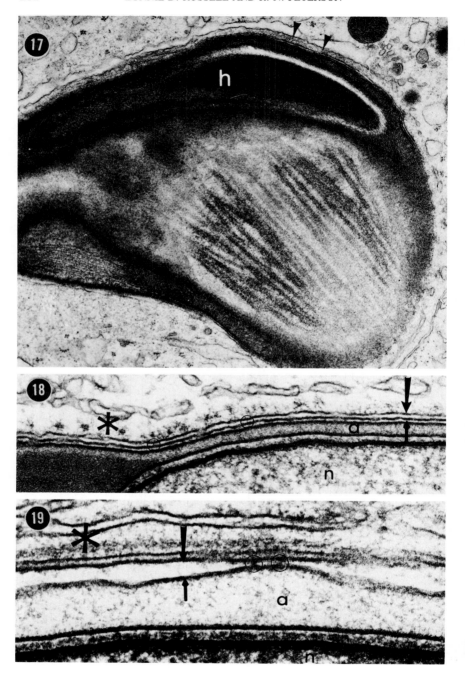

occurs is not known; several investigators (Toyama, 1976; Russell, 1977c; Gravis *et al.*, 1976; Franke *et al.*, 1978) suggest that with the demonstration of actin
(Toyama, 1976) and an ATPase (Gravis *et al.*, 1976) in this region, grasping the
head of late spermatids could occur. Attempts to demonstrate myosin in this
region have not been conclusive (see ''Note Added In Proof'' of Franke *et al.*,
1978) or have been negative (Vogl *et al.*, 1983), leaving open the question of
whether an actinomyosin system is operative. It does seem certain that the
mechanism for adhesion directly involves contacts between the two cells, as
evidenced by the trypsin sensitivity.

It is deduced from the aforementioned properties of Sertoli-spermatid ES, that
one of its functional properties is the maintenance of a tight association between
elongate germ cells and Sertoli cells. It may be suggested that the ES replaces the
desmosome in maintaining the integrity of the seminiferous epithelium. Another
functional property of ES is suggested based upon (1) the adhesiveness of Sertoli
and germ cell plasma membranes at ES sites, (2) the filament–membrane connection occuring within the ES, and (3) the connections of ES with cytoplasmic
microtubules of the Sertoli cell. As regards the last item, microtubules frequently
course within 10–30 nm of the deep aspect of the cisternae of endoplasmic
reticulum forming the ES and, in doing so, are linked by bridges to the ES
(Russell, 1977c). In the ground squirrel, tubulin is primarily localized near the
heads of late spermatids (Vogl *et al.*, 1983). We take the liberty of comparing the
properties of the ES, and associated microtubules, to those of a plunger with a
terminal suction cup. The suction cup (ES) maintains a large area of adhesion to
another surface (spermatid) while the plunger (microtubules and cytoskeletal
elements) may be used to pull or push the structure being held. In a similar way,
we believe that the ES acts as an adhesion placque or mantle which may be acted
upon by cytoskeletal structures, such as microtubules, to alter the position of a
germ cell (Russell, 1977c).

At elongation, the germ cell undergoes a fundamental change in position. It is
moved from the lateral aspect of the Sertoli cell to a position deep within crypts
or recesses within the Sertoli cell. In the rat these crypts have been termed

---

FIG. 17.   At this step of development the spermatid head (h) is flat and, for the most part, the
section is parallel to and close to the surface. The ES at the dorsal convex region of the head is shown
sectioned perpendicularly (arrowheads). The section reveals, to advantage, in some regions the
surface zone of an adjoining Sertoli cell showing the ES filament bundles sectioned longitudinally.
Bundles are generally parallel with each other and spaced about 30–50 nm apart. Bundles are
occasionally seen to anastomose with one another. ×9200.

FIGS. 18 and 19.   The relationship of the Sertoli cell to elongating spermatids is depicted. Septa
(encircled) are seen bridging the intercellular space between spermatid (short arrow) and Sertoli (long
arrow) plasma membranes. Also indicated are the spermatid nucleus (n), acrosome (a), and Sertoli
ES (asterisk). The intercellular space in Fig. 19 is not uniform, as the ES–spermatid relationship is
just being formed. ×73,000 and ×98,000, respectively.

cylindrical recesses and average about 65 μm in depth, although some of them extend almost to the basal lamina (Wong and Russell, 1983; Weber et al., 1983). All mammals demonstrate these crypts although there is considerable variability in their depth (Russell and Peterson, 1984). Translocation of germ cells into recesses of the Sertoli cell to form crypts for these germ cells and also the translocation which results in abolition of the crypts during the spermiation process is a dramatic phenomenon. In the rat the depth of the crypts changes at specific times during the spermatogenic cycle, some of the times of which are not associated with the initial formation of the crypts nor their abolition during spermiation. The ES of the fixed Sertoli cell population is ideally positioned, facing the heads of the elongate(ing) spermatids to draw them centrifugally to form Sertoli crypts and, at the appropriate time, to push them toward the lumen to initiate spermiation.

Finally, one additional function of ES may be said to occur during the spermiation process. After the late spermatid is positioned at the tubular lumen, ES undergoes a fundamental change by either dissociating and/or moving away from the head of the spermatid and/or by simply losing the adhesive property known to be associated with the plasma membranes at that site (Ross and Dobler, 1975; Ross, 1976; Russell, 1984). ES is generally not considered the final holdfast device for, as ES is lost from the spermatid head in most mammalian species, it is replaced by another holdfast device, the tubulobulbar complex (see Section IV,E). These reviewers view the loss of ES during the spermiation process as essential to the successful completion of spermiation, not only aiding to disengage the spermatid, but also to withdraw the mantle and, consequently, the Sertoli cell cytoplasm which forms a recess around the spermatid head. Temporally, the dissolution or displacement of ES accompanies the loss of the crypt or recess in which the Sertoli cell formed to accommodate the spermatid head. Thus, profiles of spermatids near release lack ES and also lack a Sertoli crypt which previously encompassed the spermatid head (Russell, 1984).

ES first appears during the mouse spermatogenic cycle apposed to mid-cycle pachytene spermatocytes (Russell et al., 1980). Several other mammalian species, incuding the rat (Russell, 1977c), examined by these reviewers display ES apposing pachytene spermatocytes at mid-cycle. Gravis (1979) showed one micrograph (his Fig. 5) of a basal compartment cell of the hamster facing ES. After extensive examination of thin sections of hamster seminiferous tubules, we suggest that Gravis' finding is atypical, for we could not find another basal compartment hamster cell, in a variety of stages examined, which showed a profile of ES like that depicted by Gravis. Ross and Dobler (1975) suggested that ES first comes to face and remain with germ cells as those germ cells cross the zone of Sertoli–Sertoli contact (Ross, 1970) where occluding junctions are present and are flanked by ES (Dym and Fawcett, 1970). The theory, no matter how appealing, was not supported by morphological documentation and is contradicted by

Brökelmann, J. (1963). *Z. Zellforsch.* **59,** 820–850.

Burgos, M. H., and Fawcett, D. W. (1955). *J. Biophys. Biochem. Cytol.* **1,** 287–300.

Camatini, M., Franchi, E., and deCurtis, I. (1979). *J. Submicrosc. Cytol.* **11,** 511–516.

Camatini, M., Franchi, E., and deCurtis, I. (1981). *Anat. Rec.* **200,** 293–297.

Camatini, M., Franchi, E., deCurtis, I., Anelli, G., and Masera, G. (1982). *Anat. Rec.* **203,** 353–363.

Cameron, D. F., and Markwald, R. R. (1981). *Am. J. Anat.* **160,** 343–358.

Cameron, D. F., and Syndle, F. E. (1980). *Fertil. Steril.* **34,** 255–258.

Cameron, D. F., Syndle, F. E., Ross, M. H., and Drylie, D. M. (1980). *Fertil. Steril.* **33,** 526–533.

Castro, A. E., and Seiguer, A. C. (1974). *Virchows Arch. Abt. B Zellpathol.* **16,** 297–309.

Cavicchia, J. C., and Burgos, M. H. (1983). *J. Cell Biol.* **97,** 17a.

Chiquoine, A. D. (1964). *Anat. Rec.* **149,** 23–35.

Clermont, Y. (1958). *Exp. Cell Res.* **15,** 438–440.

Clermont, Y., and Morales, C. (1982). *Anat. Rec.* **202,** 32–33A.

Clermont, Y., Leblond, C. P., and Messier, B. (1959). *Arch. Anat. Microsc. Morphol. Exp.* **48,** 36–55.

Clermont, Y., McCoshen, J., and Hermo, L. (1980). *Anat. Rec.* **196,** 83–99.

Connell, C. J. (1974). *Anat. Rec.* **178,** 333.

Connell, C. J. (1978). *J. Cell Biol.* **76,** 57–75.

Connell, C. J. (1980). *In* "Testicular Development, Structure, and Function" (A. Steinberger and E. Steinberger, eds.), pp. 71–78. Raven, New York.

Dym, M. (1973). *Anat. Rec.* **175,** 639–659.

Dym, M., and Cavicchia, J. C. (1977). *Biol. Reprod.* **17,** 390–403.

Dym, M., and Fawcett, D. W. (1970). *Biol. Reprod.* **3,** 308–326.

Eddy, E. M., and Kahri, A. I. (1976). *Anat. Rec.* **185,** 333–358.

Espevik, T., Lamvik, M. K., Sunde, A., and Eik-Nes, K. B. (1982). *J. Reprod. Fertil.* **65,** 489–495.

Eusebi, F., Ziparo, E., Fratamico, G., Russo, M., and Stefani, M. (1983). *Dev. Biol.* **100,** 249–255.

Farquhar, M. G., and Palade, G. E. (1963). *J. Cell Biol.* **17,** 375–412.

Fawcett, D. W. (1970). *In* "The Regulation of Mammalian Reproduction" (S. Segal, R. Gozier, P. Corfman, and P. Condiliffe, eds.), pp. 116–138. Thomas, Springfield, Illinois.

Fawcett, D. W. (1974). *In* "Male Fertility and Sterility" (R. E. Mancini and L. Martini, eds.), pp. 13–36. Academic Press, New York.

Fawcett, D. W. (1975). *Handb. Physiol.* pp. 21–53.

Fawcett, D. W., Leak, L. V., and Heideger, P. M. (1970). *J. Reprod. Fertil. (Suppl.)* **10,** 105–122.

Flickinger, C., and Fawcett, D. W. (1967). *Anat. Rec.* **158,** 207–222.

Fouquet, J. P. (1974). *J. Microsc.* **19,** 161–168.

Franchi, E., and Camatini, M. (1983). *J. Cell Biol.* **95,** 108a.

Franke, W. W., Grund, C., Fink, A., Weber, K., Jockusch, B. M., Zentgraf, H., and Osborn, M. (1978). *Biol. Cell.* **31,** 7–14.

Fritz, I. B., Lyon, M. F., and Setchell, B. P. (1983). *J. Reprod. Fertil.* **67,** 359–363.

Furuya, S., Kumamoto, Y., and Sugiyama, S. (1978). *Arch. Androl.* **1,** 211–219.

Furuya, S., Kumamoto, Y., and Ikegaki, S. (1980a). *Arch. Androl.* **5,** 361–367.

Furuya, S., Kumamoto, Y., Mori, M., and Sugiyama, S. (1980b). *In* "Normal and cryptorchid Testis" (E. S. E. Hafez, ed.), pp. 73–93. Nijhoff, The Hague.

Gilula, N. B., Fawcett, D. W., and Aoki, A. (1976). *Dev. Biol.* **50,** 142–168.

Gondos, B. (1980). *In* "Testicular Development, Structure, and Function" (A. Steinberger and E. Steinberger, eds.), pp. 3–20. Raven, New York.

Gondos, B., and Connell, C. J. (1978). *Arch. Androl.* **1,** 19–30.

Gravis, C. J. (1978). *Am. J. Anat.* **151,** 21–38.

Gravis, C. J. (1979). *Z. Mikrosk. Anat. Forsch. (Leipzig)* **93**, 321–342.
Gravis, C. J. (1980). *Tissue Cell.* **12**, 309–322.
Gravis, C. J., Chen, I. L., and Yates, R. D. (1976). *Am. J. Anat.* **147**, 419–432.
Hadziselimovic, F. (1977). *Adv. Anat. Embryol. Cell Biol.* **53**, 1–69.
Hagenäs, L., and Ritzen, E. M. (1976). *Mol. Cell. Endocrinol.* **4**, 25–34.
Hagenäs, L., Plöen, L., Ritzen, E. M., and Ekwall, H. (1977). *Andrologia* **9**, 250–254.
Hagenäs, L., Plöen, L., and Ekwall, H. (1978). *J. Endocrinol.* **76**, 87–91.
Janecki, A., Lukaszyk, A., and Jakubowiak, A. (1981). *Folia Histochem. Cytochem.* **19**, 135–142.
Johnson, M. H. (1969). *J. Reprod. Fertil.* **22**, 181–186.
Johnson, M. H. (1970a). *J. Reprod. Fertil.* **22**, 119–127.
Johnson, M. H. (1970b). *J. Reprod. Fertil.* **22**, 181–186.
Johnson, M. H., and Setchell, B. P. (1968). *J. Reprod. Fertil.* **17**, 403–406.
Kaya, M., and Harrison, R. G. (1976). *J. Anat.* **121**, 279–290.
Kelly, D. E. (1966). *J. Cell Biol.* **28**, 51–72.
Koskimies, A. I. (1973). *Ann. Med. Exp. Biol. Ferviae (Helsinki)* **51**, 74–81.
Lalli, M. F., and Clermont, Y. (1975). *Anat. Rec.* **181**, 403–404.
Larsen, W. J. (1977). *Tissue Cell* **9**, 373–394.
Leeson, T. S. (1975). *J. Morphol.* **147**, 171–186.
McGinley, D., Posalaky, Z., and Porvaznik, M. (1977). *Anat. Rec.* **189**, 211–232.
McGinley, D. M., Posalaky, Z., Porvaznik, M., and Russell, L. (1979). *Tissue Cell* **11**, 741–754.
McGinley, D. M., Posalaky, Z., and Porvaznik, M. (1981). *Tissue Cell* **13**, 337–347.
Main, S. J., and Waites, G. M. H. (1977). *J. Reprod. Fertil.* **51**, 439–450.
Martinez-Palomo, A., and Erlij, D. (1975). *Proc. Natl. Acad. Sci. U.S.A.* **72**, 4487–4491.
Meyer, R., Posalaky, Z., and McGinley, D. (1977). *J. Ultrastruct. Res.* **61**, 271–283.
Meyer, R., Posalaky, Z., and McGinley, D. (1978). *In Vitro* **14**, 916–923.
Nagano, T. (1959). *Arch. Histol. Japan.* **16**, 311–345.
Nagano, T. (1966). *Z. Zellforsch.* **73**, 89–106.
Nagano, T. (1980). *J. Electron Microsc.* **29**, 250–255.
Nagano, T., and Okumura, K. (1973). *Virchows Arch. Abt. B Zellpathol.* **14**, 223–235.
Nagano, T., and Suzuki, F. (1976a). *Cell Tissue Res.* **166**, 37–48.
Nagano, T., and Suzuki, F. (1976b). *Anat. Rec.* **185**, 403–418.
Nagano, T., and Suzuki, F. (1978). *Cell Tissue Res.* **189**, 389–401.
Nagano, T., and Suzuki, F. (1983). *Int. Rev. Cytol.* **81**, 163–190.
Nagano, T., Suzuki, F., Kitamura, Y., and Matsumoto, K. (1977). *Lab. Invest.* **36**, 8–17.
Nagano, T., Toyama, Y., and Suzuki, F. (1982). *Am. J. Anat.* **163**, 47–58.
Neaves, W. B. (1973). *J. Cell Biol.* **59**, 559–572.
Neaves, W. B. (1977). *In* "The Testis" (A. D. Johnson and W. R. Gomes, eds.), Vol. 4, pp. 125–162. Academic Press, New York.
Neaves, W. B. (1978). *J. Reprod. Fertil.* **54**, 405–411.
Nicander, L. (1967). *Z. Zellforsch.* **83**, 375–397.
Okamura, K., Lee, I. P., and Dixon, R. L. (1975). *J. Pharmacol. Exp. Ther.* **194**, 89–95.
Osman, D. I., and Plöen, L. (1978). *Int. J. Androl.* **1**, 162–179.
Pari, G. A. (1910). *Frankf. Z. Pathol.* **4**, 1–29.
Pelletier, R. M., and Friend, D. S. (1980). *J. Cell Biol.* **87**, 151a.
Pelletier, R. M., and Friend, D. S. (1983). *Am. J. Anat.* **168**, 213–228.
Ribbert, H. (1904). *Z. Allg. Physiol.* **4**, 201–214.
Romrell, L. J., and Ross, M. H. (1979). *Anat. Rec.* **193**, 23–43.
Ross, M. H. (1970). *In* "Morphological Aspects of Andrology" (A. F. Holstein and E. Horstmann, eds.), pp. 83–86. Grosse, Berlin.
Ross, M. H. (1976). *Anat. Rec.* **186**, 79–103.

Ross, M. H. (1977). *Am. J. Anat.* **148**, 49–56.

Ross, H. H., and Dobler, J. (1975). *Anat. Rec.* **183**, 267–292.

Russell, L. (1977a). *Am. J. Anat.* **148**, 301–312.

Russell, L. (1977b). *Am. J. Anat.* **148**, 313–328.

Russell, L. (1977c). *Tissue Cell* **9**, 475–498.

Russell, L. D. (1978a). *Anat. Rec.* **190**, 527.

Russell, L. D. (1978b). *Anat. Rec.* **190**, 99–112.

Russell, L. D. (1979a). *Anat. Rec.* **194**, 233–246.

Russell, L. D. (1979b). *Anat. Rec.* **194**, 213–232.

Russell, L. D. (1979c). *Am. J. Anat.* **155**, 259–279.

Russell, L. D. (1980a). *Anat. Rec.* **197**, 21–31.

Russell, L. D. (1980b). *Gamete Res.* **3**, 179–202.

Russell, L. D. (1984). *In* "Electron Microscopy in Biology and Medicine—Ultrastructure of Reproduction" (J. van Blerkhom and P. Motta, eds.), pp. 46–66. Martinus Nijhoff, Boston.

Russell, L. D., and Burguet, S. (1977). *Tissue Cell* **9**, 751–766.

Russell, L., and Clermont, Y. (1976). *Anat. Rec.* **185**, 259–278.

Russell, L. D., and Gardner, P. J. (1974). *Biol. Reprod.* **11**, 631–643.

Russell, L. D., and Karpas, S. (1983). *Anat. Rec.* **205**, 171A.

Russell, L. D., and Malone, J. P. (1980). *Tissue Cell* **12**, 263–285.

Russell, L. D., and Peterson, R. N. (1984). *J. Reprod. Fertil.* **70**, 635–641.

Russell, L., Myers, P., Ostenburg, J., and Malone, J. (1980). *In* "Testicular Development, Structure and Function" (A. Steinberger and E. Steinberger, eds.), pp. 55–63. Raven, New York.

Russell, L. D., Lee, I. P., Ettlin, R., and Peterson, R. N. (1983a). *Tissue Cell* **15**, 615–626.

Russell, L. D., Tallon-Doran, M., Weber, J. E., Wong, V., and Peterson, R. N. (1983b). *Am. J. Anat.* **167**, 181–192.

Sapsford, C. S. (1963). *J. Anat.* **97**, 225–238.

Sapsford, C. S., Rae, C. A., and Cleland, K. W. (1969). *Aust. J. Zool.* **17**, 195–292.

Setchell, B. P. (1974). *In* "Male Fertility and Sterility" (R. E. Mancini and L. Martini, eds.), pp. 37–57. Academic Press, New York.

Setchell, B. P. (1978). "The Mamalian Testis." Elik, London.

Setchell, B. P., and Waites, G. M. H. (1975). *Handb. Physiol.* pp. 143–172.

Setchell, B. P., Voglmayr, J. K., and Waites, G. M. H. (1969). *J. Physiol. (London)* **200**, 73–85.

Toyama, Y. (1976). *Anat. Rec.* **186**, 477–492.

Tung, P. S., Dorrington, J. H., and Fritz, I. B. (1975). *Proc. Natl. Acad. Sci. U.S.A.* **72**, 1838–1842.

Vitale, R. (1975). *Anat. Rec.* **181**, 501.

Vitale, R., Fawcett, D. W., and Dym, M. (1973). *Anat. Rec.* **176**, 333–344.

Vogl, W., Lin, Y. C., Dym, M., and Fawcett, D. W. (1983). *Am. J. Anat.* **168**, 83–98.

Waites, G. M. H. (1977). *In* "The Testis" (A. D. Johnson and W. R. Gomes, eds.), Vol. 4, pp. 91–123. Academic Press, New York.

Weber, J. E., Russell, L. D., Wong, V., and Peterson, R. N. (1983). *Am. J. Anat.* **167**, 163–179.

Willson, J. T., Jones, N. A., Katsh, S., and Smith, S. W. (1973). *Anat. Rec.* **176**, 85–100.

Wong, V., and Russell, L. D. (1983). *Am. J. Anat.* **167**, 143–161.

Ziparo, E., Geremia, R., Russo, M. A., and Stefanini, M. (1980). *Am. J. Anat.* **159**, 385–388.

Ziparo, E., Siracusa, G., Palombi, F., Russo, M. A., and Stefanini, M. (1981). *N.Y. Acad. Sci.* **383**, 511–512.

INTERNATIONAL REVIEW OF CYTOLOGY, VOL. 94

Functioning and Variation of Cytoplasmic Genomes: Lessons from Cytoplasmic–Nuclear Interactions Affecting Male Fertility in Plants

MAUREEN R. HANSON[*,1] AND MARY F. CONDE[†]

*Department of Biology, University of Virginia, Charlottesville, Virginia, and
†Experimental Plant Genetics, Upjohn Company, Kalamazoo, Michigan

[1]Present address: Section of Genetics and Development, Cornell University, Ithaca, New York 14853.

I. Introduction and Overview

Why the phenomenon of cytoplasmic male sterility (CMS) is so widely distributed among many different plant genera is unknown. CMS, abortion of male gametophyte development specified by cytoplasmically inherited factors, was reported by 1972 in 140 species (Laser and Lersten, 1972), and the list has grown since then. Although CMS is often a prized agricultural characteristic, human selection for the phenomenon has benefited from, not created its generality. The generality of CMS suggests that understanding its basis may give us fundamental information about cytoplasmic genomes' functioning in plant development. While other cytoplasmically controlled phenotypes are known in higher plants, namely herbicide tolerance, vegetative and floral abnormalities, disease sensitivity, antibiotic sensitivity, cytoplasmic female sterility, and tentoxin sensitivity (Durbin and Uchytil, 1977; Menzcel *et al.*, 1983; Grun, 1976; Harvey *et al.*, 1972; Machado *et al.*, 1978; Hooker, 1974), they are not as widely distributed throughout the plant kingdom as CMS. CMS therefore provides an excellent opportunity to dissect out cytoplasmic genome functions.

CMS is agriculturally valuable for the production of hybrid seed. CMS plants are used as female parents in crosses, so that any seed formed on the CMS parent is known to be the product of cross-pollination. In some crops, cytoplasmically male sterile parents are the only way to economically and effectively produce hybrid seed. The alternatives, nuclear genetic male sterility and hand-emasculation before cross pollination, are often not economically viable. There is frequently too much labor and risk involved in removing all male fertiles from a field population segregating for nuclear recessive male sterility ("rogueing-out") and not all crop plants can be easily emasculated. For example, production of hybrid seed for grain sorghum on a commercial scale is dependent on a CMS system, for the flowers are small, difficult to emasculate, and the plant is self-fertile (Section IV).

Until we understand the molecular basis of CMS, we will not know why male sterility is a common manifestation of nuclear-cytoplasmic incompatibility. Cytoplasmic genomes may have a special role in reproductive development. Alternatively, male reproductive development may be a particularly delicate developmental process, which is easily disturbed. Perhaps floral development is a rare example of a dispensable function, while other nuclear–organelle interactions are lethal or too subtle to be detected phenotypically. In some genera, CMS may have been selected for as an outbreeding device (Grun, 1976).

The finding of alloplasmic CMS—that arising from interspecific crosses—has led to the concept that CMS results from incompatibility between the nuclear genome of one species and the cytoplasm of another. Let us consider how alloplasmic CMS systems have typically been produced via interspecific crosses. A wild species is most often used as the female parent in a cross with a cultivated species. The F_1 hybrid may or may not be completely fertile. Often both male and female sterility at this point results from chromosomal and meiotic aberrations unrelated to CMS. Provided that the interspecific hybrid is not completely sterile, the cultivar is crossed as male parent in succeeding generations. As the wild species' nuclear genes are eliminated in advanced generations, CMS plants begin to appear in the population. Sufficient crosses may be performed to substitute the nuclear genome of the cultivar into the wild cytoplasmic background at the 99% level. This process has been carried out in a number of plant genera, resulting in the synthesis of defined CMS types (Laser and Lersten, 1972; Grun, 1976; Frankel and Galun, 1977).

However, CMS has arisen in intraspecific as well as interspecific crosses, spontaneously and after mutagenic treatment. This review will describe many documented instances of nonalloplasmic CMS. While the known alloplasmic nature of many CMS systems may be providing a correct clue to CMS as a nuclear-cytoplasmic incompatibility, the disruption which results in CMS may apparently also arise by genome changes within a cultivar or species. That is, perhaps alterations in cytoplasmic genetic elements within a species may also give rise to CMS by producing problems in nuclear–cytoplasmic interaction.

In most cases of CMS, nuclear genes called "restorers" of fertility have been found. Often restorers are detected by crossing nuclear genes from the original cytoplasm donor species or related species into the CMS line, but sometimes are found in different cultures or populations of the same species. Both monogenic and polygenic systems of nuclear gene CMS modification are known, and both recessive and dominant sterility modifiers have been identified. In Grun's (1976) survey of fertility restoration, 80% of CMS types arising in intraspecific crosses had recessive restorers while 40% of interspecific- or intergeneric-derived CMS types had recessive restorers.

While the wide distribution of both CMS and restorer nuclear genes may suggest a common fundamental basis for the developmental aberration—perhaps disruption in cytoplasmic-nuclear coordination—the particular mechanism of such a disruption may well vary in different CMS systems. Countless models for the molecular basis of CMS can be generated, especially by analogy to the types of mutations affecting nuclear-mitochondrial coordination and functioning in fungi, particularly yeast (see Tzagoloff, 1982; Poyton, 1983). Nuclear genes (restorer alleles) are evidently able to modulate the developmental outcome of both evolutionary divergence (interspecific cross-derived CMS) and more recent intraspecific changes (intraspecific-derived CMS) in cytoplasmic genomes.

Restorer genes are a necessity for those crops which are grown from hybrid seed generated from a CMS parent, if the desired agricultural commodity is a fruit or seed. For example, a male-sterile hybrid sunflower would not produce the desired oil seed—but a male-sterile hybrid petunia still produces the valued product, colorful flowers.

CMS-modifying restorer genes have also been, until recently, the geneticist's primary method to distinguish cytoplasms, by test crosses with defined lines. Two cytoplasms have often been considered ''identical'' if the same modifier genes affect the plants' phenotype in the same manner. Two cytoplasms are differentiated if the same nuclear modifier gene produces plants of different phenotype following crosses with two different female parents.

With the advent of molecular techniques for directly examining cytoplasmic genomes, cytoplasms can now be differentiated further. Two cytoplasms having indistinguishable reactions to modifier genes can be distinguished if their cytoplasmic genomes exhibit restriction site heterogeneity. Plastid genomes of different species have been distinguished by DNA sequence heterogeneity affecting restriction sites, although the genome is often highly conserved in both sequence and arrangement (Edelman, 1981; Lamppa and Bendich, 1979). Many plastid genomes have been mapped with respect to restriction sites. As yet only two mitochondrial genome maps have been published (Palmer and Shields, 1984; Chetrit et al., 1984). The mitochondrial genome is apparently less conserved and more heterogeneous in size between species than the chloroplast genome (Ward et al., 1981). In addition to variation in restriction sites and arrangement, perhaps further analyses of mitochondrial genomes will reveal significant differences in information content, unlike the plastid genome. The molecular biology and organization of higher plant mitochondrial DNA has recently been reviewed (Pring and Lonsdale, 1984; Lonsdale, 1984).

Although new methods for nucleic acid analysis permit a finer differentiation between organelle genomes than phenotypic tests, the problem remains to correlate functioning of cytoplasmic genetic information with aberrant reproductive development. If CMS arises from abnormal functioning of a cytoplasmic gene, how can such a gene be identified? Difficulty arises from the complexity and variability among mitochondrial genomes even of closely related species (or varieties of the same species). Simple restriction endonuclease analysis of mitochondrial genomes of CMS and fertile lines has in general revealed so many points of divergence that any genomic features relevant to CMS are masked. How can a nucleic acid sequence correlated with CMS be identified? The genetics and biology of certain CMS systems may be better suited for this task than others. Understanding the molecular basis of CMS in one system would open up new strategies for analysis for the more difficult systems, even if the particular mechanisms are not all identical. To enhance the probability of correlating the CMS phenotype with a nucleic acid sequence, the system under investigation should ideally exhibit three or more of the following characteristics: synthesis by sexual crosses, induced or spontaneous revertability, induction by mutagenic treatment, or spontaneously, graft transmissability, restoration by nuclear genes, or transmissability by somatic hybridization. Then genetic stocks can be synthesized which afford the possibility of making meaningful correlations between genetic information or gene expression and the CMS phenotype. Below the dicot and monocot systems which have three or more of the above characteristics will be considered in detail.

II. Male Sterility in *Petunia*

CMS in *Petunia* has three of the features discussed above as useful for correlating the sterile phenotype with cytoplasmic genetic information or expression. CMS in *Petunia* is graft-transmissible, modified by a number of nuclear genes, and transferable by protoplast fusion. One notable aspect of *Petunia* CMS is its stability once fixed in either graft-transmitted or somatic hybridization-transmitted progeny. Heritable CMS in *Petunia* has not been chemically, radiation, nor environmentally induced, nor is there an observable spontaneous reversion rate to fertility. In *Petunia* there is substantial, if circumstantial, evidence consistent with the hypothesis of the existence of a mobile genetic element which specifies CMS. This point will be clarified in the following discussion of genetic experiments with *Petunia* CMS and fertile lines.

A. GRAFT TRANSMISSABILITY

Of the three genera in which graft transmission of CMS has been reported, the *Petunia* system has been subject to the most rigorous genetic analyses and most intense investigation. Nevertheless, some key information and genetic material is unavailable. For example, although Everett and Gabelman deliberately synthesized CMS in *Petunia* by crossing wild species with cultivars, the species providing the CMS cytoplasm was not recorded (cited in Duvick, 1959; and Izhar, 1984). A new synthesis of CMS by an interspecific cross followed by examination of organelle genomes would be instructive. Unfortunately, we are limited to using CMS lines which are probably all derived from the original cross by Everett and Gabelman. By traditional genetic tests with restorer loci, Izhar and Frankel (1976) could not distinguish more than one CMS cytoplasm in *Petunia* CMS lines acquired from different laboratories. Restriction digests of two CMS lines, one from J.R. Edwardson (Florida) and the other from S. Izhar (Israel), produced nearly identical restriction banding patterns of mitochondrial DNA (Boeshore and Hanson, unpublished).

Frankel (1956) was the first to report graft transmission of CMS in *Petunia*. Fertile scions grafted onto a CMS stock remained fertile, but progeny of selfed flowers included CMS individuals. Further crosses with these CMS individuals produced CMS phenotypes for over 40 generations (Izhar, 1984).

Frankel's controversial report has since been verified in other laboratories. Appropriate controls have been performed to conclude that male sterile individuals do not appear in progeny of the fertile lines used unless they are grafted onto sterile (but not fertile) stocks. Edwardson and Corbett (1961) were able to obtain partially sterile individuals in the F_1 generation from fertile scions and fully CMS individuals in the F_2 generation. These workers were also able to obtain CMS progeny from fertile scions grafted onto a 12-in. tobacco segment which was itself grafted onto a sterile *Petunia* stock (Corbett and Edwardson, 1964). In 140 progeny of three selfed flowers on fertile scions, Bianchi (1963) found about 20% exhibiting some male sterility. Frankel has also performed extensive additional experiments in which he found that CMS could be graft-transmitted into progeny of flowers on fertile stocks carrying sterile scions, as well as vice versa (Frankel, 1964).

The published reports of graft transmission of CMS are difficult to compare, since no two laboratories used the same genotype for stock and scion. Within the same laboratory, there are indications that the nuclear background of the transmitting or recipient genotype may be important. Several papers indicate many unsuccessful attempts for graft transmission in addition to positive results. Only in 3 of 15 graft combinations did Bianchi (1963) find sterile progeny. Everett (cited in Edwardson and Corbett, 1961) did not achieve graft transmission of CMS. Of F_1 progeny of fertile "Heavenly Rose" scions grafted on CMS somatic hybrid plants, plants exhibited occasional sterile flowers (Hanson, unpublished). None of the F_2 progeny from selfed sterile or partially sterile flowers exhibited any male sterility. Frankel (1964) pointed out that only 4 out of 25 graft combinations he evaluated yielded some CMS F_1 progeny.

The frequency of graft transmission of CMS is often low, or not achieved until advanced generations. No completely sterile plants in the F_1 generation were observed by Edwardson and Corbett (1961). Of 1192 F_3 plants, only two sterile plants were obtained. On 9 F_1 progeny observed by Corbett and Edwardson (1964), 90 flowers were fertile, 7 were sterile, and 7 were partially sterile. In 451 F_2 plants derived from sterile or partially sterile flowers, only 252 flowers of 4914 examined exhibited sterility (Corbett and Edwardson, 1964).

One interesting aspect of CMS graft transmission is the appearance of CMS individuals in progeny of fully fertile flowers, not only on the original scion but also in the F_1 or F_2 generation. In all reports, the scion remained fertile as long as observed—for over 4 years in one case—even though progeny included sterile individuals. Fully fertile flowers on F_1 and F_2 plants also have been observed to give rise to sterile progeny. For example, Bianchi (1963) selected a fertile individual from an F_1 population containing both fertile and sterile plants, and found 20% of its progeny exhibited male sterility. However, fertile individuals in the F_1 populations evaluated by Frankel (1956, 1964) yielded only fertile progeny.

A further peculiarity of CMS in *Petunia* is Frankel's findings that the pollen for the F_1 cross on the fertile scion must be derived from either the fertile scion itself or a flower on a scion attached to another sterile stock. While other aspects of CMS graft transmission have been verified by workers using different genotypes in other laboratories, Frankel (1964) has apparently performed the only experiments testing this point, since all other reports describe the use of selfed flowers on fertile scions.

A number of factors may contribute to the variability in achievement of CMS transmission. In some cases, the scion may have carried restorer genes which masked expression of CMS. If a scion lacking restorers was used, other nuclear genes of the scion might inhibit stable transmission of a CMS factor or of a CMS-carrying entity. Alternatively, only certain stocks may have the appropriate nuclear background to transmit CMS to a fertile scion. There also may be uncontrolled environmental factors that are critical to achieve transmission. For example, Frankel (1964) comments that progeny from flowers on fertile scions selfed during the winter have higher percentages of sterile progeny. However, Frankel observed that similar frequencies of CMS graft transmission occurred in the same grafted plant over several years, and that the vegetative versus floral state of the scion at time of grafting did not affect subsequent graft transmission except that fertile flowers grafted after meiosis did not produce sterile progeny (Frankel, 1962, 1964). Perhaps there are other details of the developmental stage of stock and scion at time of grafting, or type of graft technique used, which may affect the frequency of CMS transmission.

B. Restoration of Fertility by Nuclear Genes

Both a single gene system and a multigene quantitative system of fertility restoration have been observed in *Petunia*. Edwardson and Warmke (1967) described a single dominant nuclear gene restoring fertility to lines carrying the CMS cytoplasm. Izhar (1978) found that the dominant gene restoration was not affected by temperature nor by the genetic background of the CMS line.

In the multigenic system of fertility restoration, different degrees of fertility restoration occur depending on temperature (Van Marrewijk, 1969; Izhar, 1977). The critical temperature-sensitive time in pollen development in lines carrying the multigenic restorers was found to be shortly before meiosis (Izhar, 1975). However, the stage at which breakdown of pollen development occurs in a particular genotype depends on its complement of nuclear restorer genes (Izhar, 1977). Four stable CMS lines and their complementary maintainer lines could be synthesized (Izhar, 1977) in which pollen degeneration occurred at (1) meiosis, (2) tetrads, (3) microspores, and (4) around first mitosis. In some of Izhar's (1978) crosses using multigenic restorers, there are indications that restoration is at least partially gametophytic, since higher fertility-to-sterility ratios in progeny were obtained than would be expected. In complete gametophytic restoration, the genotype of the microspore itself determines whether or not it will develop normally into a mature pollen grain, regardless of the genotype of the anther sporophytic tissue. In Izhar's (1978) crosses, pollen containing fertility restorers participated in fertilization more often than those lacking restorers.

CMS lines are not heritably genetically altered in their CMS-specifying properties by incorporation of restorers into the nuclear genome. When restorers are removed from a line carrying the CMS cytoplasm, the CMS phenotype again appears.

C. Transmission of CMS by Protoplast Fusion

New combinations of cytoplasmic and nuclear genotypes in *Petunia* via protoplast fusion have been achieved in several laboratories. Bergounioux-Bunisset and Perennes (1980) fused protoplasts of a fertile line and a CMS line and obtained 3 male fertile plants with nuclear-controlled floral color phenotypes characteristic of the sterile line. Six male fertile somatic hybrids were also obtained. Following fusion of protoplasts of a CMS line with protoplasts of fertile *P. axillaris,* Izhar and Power (1979) regenerated more than 140 fertile plants which had the phenotype of *P. axillaris.* In the selfed progeny of 6 of the regenerated plants, CMS individuals appeared. A restorer allele was later found in the *P. axillaris* line, a finding which could explain why CMS was not expressed in the original regenerates (Izhar, 1984).

Further CMS-fertile somatic hybridizations performed by Izhar and Tabib (1980) and Izhar *et al.* (1983) gave results reminiscent of the segregation of CMS in advanced generations following graft transmission. For example, following fusion of fertile *P. axillaris* and a CMS line, a fertile F_1 plant carrying the nuclear genome of the sterile line was obtained. In the selfed progeny of this cybrid were two sterile and 27 fertile plants. Of the 27 fertile F_2 plants 13 gave only fertile F_3 progeny, but 14 segregated both sterile and fertile plants. Thus CMS progeny can appear after two meiotic cycles in both transmission by grafting and by protoplast fusion.

In what may be the most massive somatic hybridization experiment yet undertaken, Izhar and his colleagues (1983) have further defined the segregation of CMS after protoplast fusion. Following somatic hybridization of a CMS and fertile line lacking restorers, 3923 fertile somatic hybrid plants were obtained. A sampling of 77 selfed populations from these plants yielded all fertile progeny. However, also observed were 92 stably sterile plants, from which 40 testcross populations yielded only CMS progeny. In addition, 32 plants were identified which yielded both CMS and sterile progeny. Among these 32 plants were seven which had both sterile and fertile branches. All progeny from sterile flowers were CMS and all from fertile branches were fertile. Since these somatic hybrid plants could be viewed as analogous to CMS stock-fertile scion graft combinations, the absence of CMS progeny from fertile branches is notable. One possible explanation is an unsuitable nuclear background to permit graft transmission.

Other plants among the 32 yielding both sterile and fertile progeny had branches which converted from CMS to fertile over several months growth, or which had flowers with two or three anthers fertile and the others sterile. These plants, as well as two others which had only sterile flowers, all gave rise to both fertile and CMS progeny. Some of the CMS progeny continued to segregate male fertile individuals for two more generations (Izhar *et al.*, 1983). Thus both CMS and fertile plants can yield both types of progeny following somatic hybridization, while in graft transmission, only fertile scions and fertile progeny have been observed to segregate both types of progeny.

To explain the somatic and gametic segregation of fertility and sterility, Izhar *et al.* (1983) have postulated sterility ''ste'' elements and fertility ''fte'' elements. These elements could be organelles or other cytoplasmic entities which sort out both in somatic and gametic tissue. Since gametic segregation of CMS and fertility occurred in tissue which did not somatically segregate for CMS and fertility, Izhar *et al.* (1983) have suggested that the gametic sorting out is more rapid than somatic sorting out. A mixed population of cytoplasmic entities, such as organelles, which are present in lower number in egg cells than in somatic cells, would be theoretically expected to segregate more rapidly after constriction to lower population size (Michaelis, 1957).

An ''ste'' element segregating in somatic hybrids need not necessarily be the same as the sterility factor transmitted through grafting. As just one example, the ''ste'' and ''fte'' elements in somatic hybrid tissue could be mitochondria from sterile and fertile parental protoplasts, while the sterility factor transmitted by grafting could be a nucleic acid which ''transforms'' some of the fertile scion mitochondria to a CMS-specifying state.

D. Molecular Analyses of Genomes in CMS and Fertile Lines

One restriction fragment difference has been detected between the chloroplast DNAs of the CMS and fertile lines used in the fusion experiments of Izhar and Power (1977) and Izhar *et al.* (1983). Use of this chloroplast genome marker allows determination of whether or not CMS assorts independently of chloroplast genomes in these hybrids. The chloroplast genome of the fertile parent has been found to be present in some of the sterile somatic hybrid plants of Izhar *et al.* (Clark *et al.*, 1985). This observation is evidence that the chloroplast genome is not specifying CMS in *Petunia*.

Numerous differences in restriction fragment patterns have been identified between mitochondrial genomes from CMS and fertile lines (Boeshore *et al.*, 1983). The configuration of the large (over 300 kb) *Petunia* mitochondrial genomes is currently unknown; both long linear DNA molecules and small (1–20 kb) circular molecules have been observed in the electron microscope (Kool *et al.*, 1982; I. Lifshitz and J. Rosinski, unpublished data).

Comparisons of mitochondrial DNA from CMS and fertile lines have not revealed any "plasmid-like" or other molecules which can be definitely correlated with the sterile or fertile genotype. Any correlation of the mitochondrial genome from the CMS parent with the mitochondrial genomes in CMS somatic hybrids is complicated by the finding that none of 12 somatic hybrids' mitochondrial genomes are identical to either parent. All *Petunia* somatic hybrid mitochondrial genomes analyzed have been found to have new combinations of restriction fragments from both parents (Boeshore *et al.*, 1983; Fig. 1).

The observation that somatic hybridization results in "hybrid" mitochondrial genomes is a great advantage in attempts to identify any CMS-specific mitochondrial DNA sequences which may exist. Somatic hybridization provides a reservoir of diverse mitochondrial genomes in CMS and fertile plants, allowing any putative CMS sequences to be readily tested for presence, arrangement, and expression in CMS vs fertile lines. For example, the mitochondrial genomes of a large number of sterile and fertile somatic hybrids can be examined for digestion with a diversity of restriction enzymes in order to determine whether any specific fragments are invariably associated with either the sterile or fertile phenotype (Hanson, 1984). Thus far, one mtDNA arrangement unique to the *Petunia* CMS parent has been shown to be present in all 17 stable-sterile somatic hybrids tested and absent from all 24 stable-fertile somatic hybrids examined (Boeshore *et al.*, 1984).

E. Lack of Reversion, Induction, Transmission through Pollen

Neither reversion of CMS in *Petunia* nor transmission of CMS through pollen can be achieved at will, and induction is limited to transmission by grafting and

Blackhull kafir as the male parent (Stephens and Holland, 1954). Selections using Texas Blackhull kafir as the recurrent male parent always produced plants which exhibited CMS; no evidence of female sterility was observed. Dwarf Yellow Sooner plants themselves are fertile; this line apparently carries a restorer gene or genes for the milo cytoplasm.

B. Molecular Analyses

Restriction endonuclease analysis indicated that milo mtDNA and ctDNA could readily be distinguished from that of the kafir lines (Pring et al., 1982a); differences between mtDNAs of these two sorghum lines were greater than those observed for ctDNAs. EcoRI digests of kafir and milo ctDNA also clearly resembled the EcoRI pattern of ctDNA from Zea mays; BamHI digests of the sorghum and maize ctDNAs did not indicate this resemblance. This suggested that there may be conservation of EcoRI sites between related species (Pring et al., 1980b).

Hybrid grain sorghum seed production has relied on the milo CMS cytoplasm almost exclusively; as a result of maternal inheritance there is homogeneity with respect to the cytoplasmic components of all hybrids. Several potential alternative sorghum CMS cytoplasms have been reported (Rao, 1962; Webster and Singh, 1964; Ross and Hackerott, 1972; Nagur and Menon, 1974; Schertz, 1977; Schertz and Ritchey, 1978). Restriction endonuclease analysis of the mtDNAs and ctDNAs of a number of these alternative sources indicated mtDNA was more variable than ctDNA among them (Pring et al., 1982a) and further that in six cases mtDNA restriction patterns were correlated with observed differences in fertility restoration (Conde et al., 1982). This provides circumstantial evidence that sorghum mtDNA rather than ctDNA is involved in the expression of CMS. Furthermore, these observations as well as those of others (Appadurai and Ponnaiya, 1967; Kidd, 1961; Maunder and Pickett, 1959) suggested that there may be more than one possible nuclear fertility restorer gene for sorghum since lines which restored some of the CMS cytoplasms did not necessarily restore all lines tested.

One alternative sorghum CMS cytoplasm (IS1112C) was characterized by the presence of two linear plasmid-like DNAs, N-1 and N-2, of 5.7 and 5.3 kbp, respectively. The IS1112C mtDNA also contained a 2.2 kbp molecule reminiscent of the 2.3 kbp circular molecule in N, C, and S maize mtDNA and up to five lower molecular weight DNAs ranging in size from 1.0 to 1.4 kbp (Pring et al., 1982b). Unlike S-1/S-2 in maize, N-1 and N-2 do not appear to share homology with any portion of the remainder of the IS1112C mtDNA; however, transcripts of N-1 and N-2 have been identified (Chase and Pring, personal communication). Unlike the maize situation in which nuclear genes affect the relative amount of S-1 and S-2 mtDNAs, no similar effect could be determined in examining 43 different sorghum lines, each carrying the IS1112C cytoplasm with N-1 and N-2 mtDNA molecules. To date there exists no genetic or molecular evidence in sorghum which would implicate these molecules in the control of CMS expression in a manner analogous to S-1 and S-2 in maize.

As in maize CMS sources, [^{35}S]methionine-labeled polypeptides synthesized by isolated mitochondria from sorghum are specific indicators for CMS (Dixon and Leaver, 1982). Mitochondria from milo CMS cytoplasm synthesized a 65-kDa polypeptide not characteristic of kafir (or fertile) cytoplasm. A number of alternative sorghum CMS cytoplasms were tested and, with the exception of three, produced the same set of polypeptides as did the milo cytoplasm. The fact that a number of mtDNAs from alternative CMS sorghum cytoplasms have been shown to have sequences which differ from that of milo (Pring et al., 1982a) and yet synthesize the same polypeptides as do mitochondria isolated from milo is reminiscent of the observation that mtDNA sequence diversity has been observed in maize N and C cytoplasm but that this is not correlated with altered polypeptides (Forde et al., 1980; Levings and Pring, 1977; Pring et al., 1980a).

The three exceptional CMS cytoplasms all produced some specific polypeptides which differed from those of both milo and kafir. IS1112C, which carries the linear N-1 and N-2 molecules (Pring et al., 1982b) was characterized by a lack of the milo-specific 65-kDa polypeptide, the presence of several high-molecular-weight polypeptides (54,000–82,000), and a 12-kDa polypeptide. M35-1 mitochondria produced patterns similar to that of IS1112C mitochondria, but lacked the 12-kDa polypeptide. Mitochondria from 9E sorghum cytoplasm were missing a 38-kDa polypeptide (characteristic of all other cytoplasm sources examined) and produced a new 42-kDa polypeptide. The 9E-specific 42-kDa polypeptide and the 38-kDa polypeptide were shown to be related through immunoprecipitation by an antiserum against yeast cytochrome oxidase subunit I. Examining genomic sequences for cytochrome oxidase subunit I may reveal why different products are made in vitro. Correlating such differences with the CMS phenotype, however, may require revertants or recombinant somatic hybrid genomes.

V. Male Sterility in Sugar Beet

A. Graft Transmission and Spontaneous Occurrence

Curtis (1967) has provided the only report of graft transmission of CMS in sugar beet. While in *Petunia*, fertile scions remained fertile, in sugar beet they either remained fertile, became partially male sterile, or completely sterile. Flowers on scions of all three phenotypes all gave rise to at least some CMS

Species	Common name	No. Hb fractions (HbC)	No. Hb fractions	Qualitative ontogeny	Quantitative ontogeny	Reference
Salmo kamloops	Kamloops Trout	6(3)	6	—	—	Yoshiyasu (1973a)
		6(3)	5	—	—	Iuchi (1973)
		6(3)	6	—	Yes	Yamanaka et al. (1967)
Salmo salar	Atlantic Salmon	—	—	—	—	—
		8	5	Yes	Yes	Koch et al. (1964, 1966, 1967)
		7	9	—	Yes	Wilkins (1968, 1970)
		8	5	Yes	Yes	Westman (1970)
Salmo trutta	European Trout	—	—	Yes	Yes	Koch et al. (1966)
		5	5	—	—	Yamanaka et al. (1967)
		8	3	Yes	Yes	This paper
Salvelinus alpinus	Arctic Char	—	—	—	—	Tsuyuki and Gadd (1963)
Salvelinus fontinalis	Brook Trout	8	7	—	Yes	Yamanaka et al. (1967)
		6(1)	8	—	—	Yoshiyasu (1973b)
		6	8	Yes	Yes	Iuchi et al. (1975)
Salvelinus leucomaenis	Japanese Char	5	7	—	—	Yamanaka et al. (1967)
		6(1)	5	—	—	Yoshiyasu (1973a)
		6(3)	5	—	—	Yoshiyasu (1973b)
		6	5	—	—	Yoshiyasu and Humoto (1972)
Salvelinus malma	Dolly Varden	6(1)	8	—	—	Yamanaka et al. (1967)
		6	10	—	—	Yoshiyasu (1973b)
Salvelinus namaycush	Lake Trout	—	—	—	—	—
Salvelinus salvelinus	Alpine Char	—	—	—	—	—

[a] List of salmonid species indicating the number of hemoglobin fractions observed and the occurrence of quantitative and qualitative ontogenetic variation in their expression. Numbers in parentheses in the HbC column indicate the number of HbC fractions which migrate to the anode (see text). Quantitative ontogeny refers to changes in the HbA/HbC ratio; qualitative ontogeny indicates changes in the occurrence of hemoglobin fractions which are not due simply to changes in the HbA/HbC ratio. A dash (—) indicates that no data are available.

to the cathode; they will hereafter be termed Hb A and Hb C, respectively. They were also shown not to be single, homogeneous components, but to consist of families or groups of hemoglobins comprising up to a total of about 22 molecular species. These detailed multiple hemoglobins will be considered later; at this point, only the relative proportions of the total A group and the total C group will be considered.

Vanstone *et al.* (1964), in a study of coho salmon (*O. kisutch*) adults and fry, confirmed that a reduction in the Hb A/Hb C ratio occurred in this species also as the fish grew in length. They showed that this change in ratio involved a real increase in the amount of Hb C. No cathodal hemoglobins were observed in coho fry. The onset of smoltification and seaward migration resulted in some increase in certain anodal hemoglobins and the first appearance of cathodally migrating hemoglobin. The amount of cathodal hemoglobin continued to increase after the fish entered sea water, and the typical adult proportions of Hb A and Hb C were attained after 6 months in the sea. The hemoglobin phenotype of large, 1-year-old migrating smolts was more like that of adults, while the phenotype of small migrating smolts was more like that of fry.

Similar changes in the Hb A/Hb C ratio were observed in sockeye salmon (*O. nerka*) and in its landlocked variant, the kokanee (Vanstone *et al.*, 1964). The fry exhibited some cathodal hemoglobin even at the smallest size examined and this increased in amount in the smolting and postsmolting period both in sockeye in the ocean and in kokanee in the lake, so that the Hb A/Hb C ratio declined steadily throughout this period. The results of Vanstone *et al.* (1964), then, indicate clearly that the decline in the Hb A/Hb C ratio (1) is due, in part, to a real increase in the *amount* of cathodally migrating hemoglobin, (2) is not simply a response triggered by the change in osmotic medium on entry to sea water, and (3) is not a function of smoltification directly, but is correlated with increasing size of fish.

Changes in the Hb A/Hb C ratio in Atlantic salmon *S. salar* and in European trout *S. trutta* have been studied in detail by Koch *et al.* (1966), Koch (1982), and Westman (1970). Koch *et al.* (1966) showed that in salmon of approximately 5 cm length, anodally migrating hemoglobin constituted 75% of the total. The proportion was even higher (82%) in a single specimen of 4 cm length, and it was suggested that in even smaller individuals the proportion of Hb A reached 100% asymptotically, with no cathodal hemoglobin being expressed in fish of less than about 3 cm. As the fish grew in length, the Hb A/Hb C ratio declined gradually so that Hb A constituted only about 60% in individuals of 20 cm length. A slower, but steady, reduction was observed in larger fish, although the Hb A/Hb C ratio did not fall below 50% even in fish over 90 cm in length. It is clear from the actual gels and the densitometric traces in these papers that the change in proportions of Hb A and Hb C is due mainly to a real increase in the amount of cathodally migrating hemoglobin.

No significant differences were observed between wild and hatchery reared individuals of normal sea-going stocks. However, relatively high values of Hb A were recorded in specimens from landlocked salmon populations. In addition, unusually high values of Hb A for their size were also recorded in a few small specimens bearing injuries in fresh water and in large, sexually maturing, specimens caught in the sea and infested with a large number of ectoparasitic copepods (*Lepeophtherus salmonis*). Individuals fed on diets rich in thyroid tissue exhibited Hb A/Hb C ratios characteristic of their actual length, i.e., thyroid in the diet had no effect on the Hb A/Hb C ratio.

Westman (1970) confirmed and extended these observations on both sea-going and landlocked *S. salar*. He showed that during growth from 3.5 to 10 cm, the proportion of Hb A decreased in both kinds of salmon from about 71 to about 60%. The proportion decreased to about 52% during subsequent growth to 90 cm in sea-going individuals. Westman attributed the change in proportion mainly to changes within the anodal group of hemoglobins together with a real increase in the amount of cathodal hemoglobin. A very significant increase in the amount of cathodal hemoglobin can be confirmed by making calculations from the densitometric traces in Fig. 2 of Westman (1970), but these fail to show any significant decrease in the amount of anodal hemoglobin.

This point is important: the decline in the Hb A/Hb C ratio signifies a major *increase* in Hb C rather than a major decrease in Hb A.

Of particular interest was an unexpected increase from 58 to 65% ($p < 0.001$) in the proportion of Hb A in landlocked salmon in the length range 20–28 cm. During subsequent growth the proportion decreased again to values nearer normal. In sea-going salmon a similar, but very much smaller, reversal was observed during growth from 17 to 31 cm. Koch (1982) showed that when smolts of a normal sea-going stock were retained in fresh water they, too, exhibited a higher proportion of Hb A than siblings released to sea. Combining these observations with those on the effect of injuries and parasitic infestation, it seems reasonable to conclude that factors which cause stress to the fish may have a measurable, but reversible, effect on the Hb A/Hb C ratio.

A decline in the Hb A/Hb C ratio with increasing length occurs also in European trout *S. trutta* (Koch *et al.*, 1966; Westman 1970) and in coho salmon *O. kisutch*. In the latter species, hemoglobins A6–A8, which correspond to Hb A in other salmonids, comprise 100% of the hemoglobin in 3-month-old fry (Giles and Vanstone, 1976). This proportion decreased in older fish due mainly to the appearance and increase in the amount of new cathodal hemoglobins. In 13-month-old presmolts, for example, Hb A made up about 85% of the total hemoglobin and in 36-month-old, sexually mature individuals, about 55%.

In comparing the data on these *Oncorhynchus* and *Salmo* species, it is noteworthy that even in the largest individuals the proportion of Hb A never declines below a minimum value of 50–55% of the total hemoglobin.

In a small number of rainbow trout *S. gairdneri* analyzed by Yamanaka *et al.* (1967), the ratio of Hb A to Hb C appeared higher in O-group individuals than in those of 1, 2, 3, and 4 years old. This observation has not been clearly demonstrated by other authors working with this well-studied species. No variation in Hb A/Hb C ratio was reported in the brook trout *Salvelinus fontinalis* examined by Yamanaka *et al.* (1967).

In summary then, all salmonid species examined exhibit multiple hemoglobins which separate on electrophoresis into distinct anodally migrating and cathodally migrating components or families of components. In all the species adequately examined (chum, coho, sockeye, and Atlantic salmon, European trout, and rainbow trout), except the brook trout, the ratio of Hb A to Hb C declines as the fish grows in length. Hb A constitutes almost 100% of the hemoglobin in small (< 5 cm) individuals, and declines to 50–55% in large, sexually mature individuals. This change is not directly attributable to, but is influenced by, smoltification, seaward migration, and environmental conditions. The change may be temporarily halted, or even reversed, by conditions of stress. The change involves a real increase in the amount of cathodally migrating hemoglobin; a real reduction in the amount of Hb A has never been unequivocally demonstrated in any species. Finally, Hb A never falls below 50% of the total hemoglobin present.

IV. Qualitative Ontogenetic Variation

With the increasing use of starch gel electrophoresis, the two hemoglobins Hb A and Hb C were seen to resolve into two separate groups, or families, of components. Generally, all the members of the Hb A group migrated electrophoretically to the anode, and all those of the Hb C group to the cathode. In a number of cases (see Table I) some components which migrated to the anode appeared to belong logically to the Hb C group. Even in these cases it was possible to assign correctly each of the components to either the Hb A or Hb C group. Ronald and Tsuyuki (1971), for example, showed by subunit analysis that one anodally migrating hemoglobin in rainbow trout was, in fact, a member of the Hb C group.

The number and precise electrophoretic mobility of the components observed in each group are, of course, influenced by the pH, ionic strength, and other physical characteristics of the starch gel medium (Koch *et al.*, 1967) and by the oxidation state of the heme iron (Tsuyuki and Ronald, 1971). Such differences account for many of the apparently conflicting results of different authors as listed in Table I. Hemolysates are, therefore, often converted to the more stable cyanmet- or carboxy-hemoglobin state prior to analysis. The multiplicity of

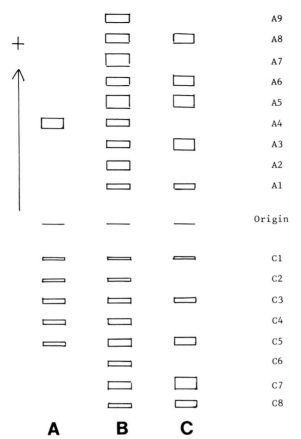

FIG. 4.   Diagram of the electrophoretic patterns of the hemoglobins of Atlantic salmon (*S. salar*): (A) small fish, ∼ 5 cm; (B) large fish, ∼ 50–60 cm; (C) very large, sexually mature fish > 80 cm. The patterns are a composite of the anodal hemoglobins as separated by Wilkins (1968) and the cathodal hemoglobins as separated by Koch *et al.* (1964) (original).

In another extensive study of Atlantic salmon hemoglobin carried out by Westman (1970), the results of Koch *et al.* (1964) on the cathodal hemoglobins were thoroughly confirmed and the pattern changes were correlated closely with growth over the range 5 to 80 cm in both sea-going and landlocked populations. The ontogeny was confirmed to occur within individual fish by repeated sampling of marked individuals as they grew over periods up to 3 years.

The polypeptide subunit structures of the anodal hemoglobins of *Salmo salar* were determined by Wilkins (1970). Four distinct polypeptide chains were detected in acid urea gels. The four polypeptides migrated electrophoretically as two pairs of components, one pair being labeled R,S and the other T,V. The apparent MW of each polypeptide was about 15,400, and of each tetramer, 56,000. Each anodal hemoglobin was a tetramer composed of one, or both, of the R,S pair together with one, or both, of the T,V pair (Fig. 5). Nine distinct tetramers of this structure are possible and all of these were observed as anodal hemoglobins: RRTT (Hb A9), RRTV (Hb A7), RRVV (Hb A4), RSTT (Hb A8), RSTV (Hb A5), RSVV (Hb A2), SSTT (Hb A6), SSTV (Hb A3), SSVV (Hb A1)—see Fig. 5. Each hemoglobin had equal amounts of the R,S and T,V pairs and asymmetric molecules of the type RV_3 or RS_2V did not occur. By analogy with sockeye salmon (see above) the subunits R,S and T,V can, for convenience, be labeled α^{Fs}, α^{Ff} and β^{Fs}, β^{Ff}, respectively (as in Fig. 5). However, it should be emphasized that this terminology is applied only in accordance with the usage whereby the components with the slower cathodal mobilities in acid starch gels are designated α and the others, β. There are still no sequence data or other evidence to confirm that salmonid α and β chains are homologous with the corresponding chains of mammalian hemoglobin.

Knowing the proportion and the types of polypeptide chains which constitute each anodal hemoglobin and the mean relative concentration of each hemoglobin in fish of different length (Wilkins, 1968) it was possible to calculate the relative proportions of the different polypeptide subunits as the fish grew in length (Fig. 6). The proportions of $R(\alpha^{Fs})$ and $V(\beta^{Ff})$, the only polypeptides present in the

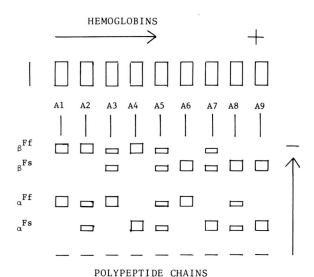

Fig. 5.   The subunit structures of the anodal hemoglobins of the Atlantic salmon. (Original, after Wilkins, 1970.)

single anodal hemoglobin A4 of very small salmon, are progressively reduced and those of $S(\alpha^{Ff})$ and $T(\beta^{Fs})$ correspondingly increased as the fish grows. Thus, the qualitative ontogenetic development of the anodal hemoglobins, involving an initial increase and later reduction in the number of hemoglobins, reflects a gradual unidirectional shift in the occurrence of polypeptide subunits, $S(\alpha^{Ff})$ replacing $R(\alpha^{Fs})$ and $T(\beta^{Fs})$ replacing $V(\beta^{Ff})$ as the fish grows bigger. This replacement, or switching, process occurs at different rates within the R,S and T,V polypeptide pairs (Fig. 6).

The structures of the cathodal hemoglobins of the Atlantic salmon have not yet been fully determined. They certainly have no chains in common with the anodal hemoglobins, although their general structure is similar. Wilkins (1970) observed three distinct subunits viz. $M(\alpha^{Ss})$, $P(\alpha^{Sf})$, and $W(\beta^S)$ and inferred that at least one more subunit, not adequately resolved in the acid-urea gel system, must exist in order to explain the 8 cathodal hemoglobins.

In the European trout S. trutta, a species closely related to the Atlantic salmon, there is only a single anodal hemoglobin present in juveniles. With increase in length two further anodal hemoglobins appear so that only three major anodal fractions are present in large individuals. Up to 8 cathodal hemoglobins which also exhibit ontogenetic variations are observed, but these have not been fully investigated yet. The subunit structures of the anodal haemoglobins are composed of a single α^F chain and two β^F chains. The single anodal hemoglobin of small individuals is $\alpha_2^F\beta_2^{Ff}$. The ontogeny involves the gradual appearance of the β^{Fs} chain so that three tetramers $\alpha_2^F\beta_2^{Ff}$, $\alpha_2^F\beta^{Ff}\beta^{Fs}$, and $\alpha_2^F\beta_2^{Fs}$ exist in the large

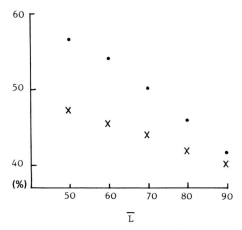

FIG. 6.   Changes in the proportions of the subunit chains of the anodal hemoglobins of Atlantic salmon as the fish grows in length. (Redrawn with permission from Wilkins, 1970.) (●) R as a percentage of $(R + S) = [\alpha^{Fs}/(\alpha^{Ff} + \alpha^{Fs})]$; ($\times$) V as a percentage of $(V + T) = [\beta^{Ff}/(\beta^{Ff} + \beta^{Fs})]$.

fish (Wilkins, unpublished). The absence of a second α^F chain explains why the anodal pattern of this species is so much more simple than that of its close congener *Salmo salar*. There are four chains, α^{Sf}, α^{Ss}, β^{Sf}, and β^{Ss}, making up the cathodal hemoglobins.

Ronald and Tsuyuki (1971) analyzed the subunit structures of the hemoglobins of the rainbow trout and the coastal cutthroat trout. Although neither species exhibited ontogenetic variations, the structures were of the familiar salmonid type involving α^F, β^F, α^S, and β^S subunits. The three anodal hemoglobins of rainbow trout possessed a common α^F subunit. There were two variants of the β^F chain and the three anodal hemoglobins had the following structures: A5, $\alpha_2^F\beta_2^{Ff}$; A6, $\alpha_2^F\beta^{Ff}\beta^{Fs}$; A7, $\alpha_2^F\beta_2^{Fs}$.

There were three major hemoglobins, one of which migrated anodally, in the Hb C group of the rainbow trout. Each Hb C hemoglobin had two α^S chains, α^{Sf} and α^{Ss}. There were also two β^S chains, β^{Sf} and β^{Ss} and the Hb C hemoglobins had the structures C1, $\alpha^{Sf}\alpha^{Ss}\beta_2^{Sf}$; C3, $\alpha^{Sf}\alpha^{Ss}\beta^{Sf}\beta^{Ss}$; A4, $\alpha^{Sf}\alpha^{Ss}\beta_2^{Ss}$. The subunit structures of the rainbow trout hemoglobins are illustrated in Fig. 7. The absence of molecules incorporating α_2^{Sf} and α_2^{Ss} subunit combinations was not explained by Ronald and Tsuyuki (1971).

The three anodal hemoglobins of the coastal cutthroat trout had only one kind

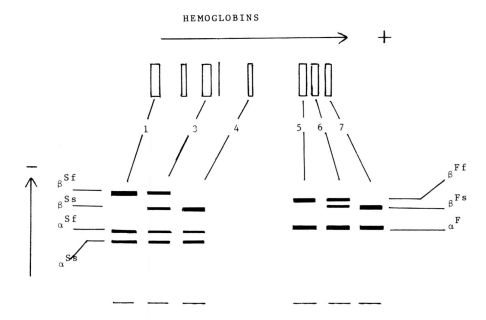

FIG. 7.   The subunit structures of the hemoglobins of rainbow trout (*S. gairdneri*). (Redrawn with permission from Ronald and Tsuyuki, 1971.)

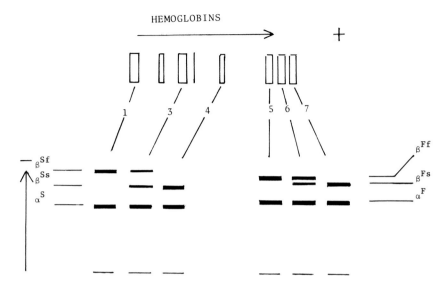

POLYPEPTIDE CHAINS

FIG. 8. The subunit structures of the hemoglobins of coastal cutthroat trout (*S. clarkii*). (Redrawn with permission from Ronald and Tsuyuki, 1971.)

of α^F chain which was common to all. There were two β^F chain variants and the formulas of the anodal hemoglobins were A5, $\alpha_2^F\beta_2^{Ff}$; A6, $\alpha_2^F\beta^{Ff}\beta^{Fs}$; A7, $\alpha_2^F\beta_2^{Fs}$. The cathodal hemoglobins were similar, with a single form of α^S chain and two variants of the β^S chain as follows: C1, $\alpha_2^S\beta_2^{Sf}$; C3, $\alpha_2^S\beta^{Sf}\beta^{Ss}$; A4, $\alpha_2^S\beta_2^{Ss}$. The cutthroat trout hemoglobins are illustrated in Fig. 8. Table II summarizes the subunit structure data for the various salmonid species.

TABLE II

NUMBER OF ELECTROPHORETICALLY DISTINCT SUBUNIT CHAINS IN THE POSTEMBRYONIC HEMOGLOBINS OF SALMONIDS

Species	Number of polypeptide chains			
	α^S	β^S	α^F	β^F
Salmo clarkii	1	2	1	2
Salmo gairdneri	2	2	1	2
Salmo trutta	2[a]	2[a]	1	2[a]
Salmo salar	2[a]	2[a]	2[a]	2[a]
Oncorhynchus nerka	2	2[a]	1	3[a]

[a]The expression of the chains within a group shows ontogenetic regulation.

V. Embryonic Hemoglobins

In the cases reviewed so far, only the postembryonic stages of the fishes' life cycles have been investigated and the sites of erythropoiesis in these are obviously independent of egg or yolk sac structures. However, studies on rainbow trout *S. gairdneri* (Iuchi, 1973; Yamamoto and Iuchi, 1975) and coho salmon *O. kisutch* (Giles and Vanstone, 1976) present evidence of the occurrence of true embryonic hemoglobins in the earliest part of the life cycle of these species.

Iuchi (1973) examined rainbow trout alevins (which still had the yolk sac attached) within 1 day of hatching and compared their hemoglobin with that of 2-year-old fish of 15–20 cm length. The embryonic hemoglobin was characterized by a higher oxygen affinity and lower Bohr effect than adult hemoglobin. In starch gel electrophoresis, freshly drawn embryonic hemoglobin was separated into 9 anodal components. Under similar conditions adult hemoglobin separated into two families of hemoglobin, the Hb A family comprising 5 anodal fractions and the Hb C family comprising 6 components, 3 of which migrated to the anode. (The C family was reduced to 3 components when 0.1 M mercaptoethanol was present in the starch gel. These latter three components may represent components 1, 3, and 4 of Ronald and Tsuyuki, 1971.) Iuchi stated that "none of the adult haemoglobin components seems to correspond to any of the larval [embryonic] haemoglobin components" although this is difficult to substantiate in the gels illustrated in Iuchi (1973). Electrophoresis of the globin subunits in acid-urea gels at pH 3.3 indicated that the embryonic hemoglobins consisted of at least some subunits distinct from those of adults; at pH 2.8 the embryonic and adult globins were regarded as quite distinct. Thus, the embryonic and postembryonic hemoglobins differ in their structure, O_2-carrying characteristics, and electrophoretic phenotype, and hemoglobin ontogeny must involve the disappearance, early in life, of the embryonic components and their complete replacement by the adult hemoglobins. These changes are fully in accord with erythropoietic events in the embryo and fry. Iuchi (1973) and Yamamoto and Iuchi (1975) have indicated that prehatch embryos synthesize a primitive red cell line (L-line) in the yolk sac. These red cells synthesize embryonic hemoglobin and persist for some time in the circulation of posthatch fry. As the population of the L-line cells declines it is replaced by the cells of the definitive line (A-line) which in the early fry stage synthesize juvenile hemoglobins, and in the later stages, adult hemoglobins. The kidney and spleen are the sites of synthesis of the A-line cells.

Giles and Vanstone (1976) observed similar results in an electrophoretic study of prehatch embryos, posthatch alevins, fry and adults of the coho salmon, *O. kisutch*. Twelve closely spaced anodal bands and one diffuse cathodal band were observed in prehatch embryos and in alevins up to 15 days posthatch (Fig. 2). Over the period 15–42 days posthatch, a reduction in the relative concentration of all the components except A6–8 was observed. By this time the alevins had absorbed most of their yolk and were emerging from the gravel. At 14 weeks posthatch, and continuing for approximately 10.5 months, hemoglobins A6–A8 comprised over 95% of the total Hb, the remainder consisting of traces of some cathodal components. Approximately 1 year after hatching, hemoglobins A1–A3 and C1–C5 made progressively greater appearance so that by 2 years posthatch the adult pattern of three major Hb A hemoglobins (A6, A7, A8) and 7 Hb C hemoglobins (A1, A3, and C1–C5) was established (Fig. 2). The Hb A hemoglobins constituted about 60% of the final adult hemoglobin and the Hb C, 40%. The electrophoretic results of Giles and Vanstone (1976) and Iuchi (1973) are remarkably similar. The timing of the disappearance of the embryonic hemoglobins in coho salmon suggests that they, like the embryonic haemoglobins of rainbow trout, are synthesized in L-line cells formed in the yolk sac. By further analogy with rainbow trout (Iuchi, 1973), the A6–A8 hemoglobins may represent the products of the immature A-line which on eventual maturation to the A-line, produces all the adult hemoglobins. The retention of three embryonic hemoglobins to adult life in the coho salmon contrasts with Iuchi's observations on rainbow trout. But closer inspection of Iuchi's illustrations and his results in acid-urea electrophoresis at pH 3.3 suggests that some, at least, of the trout anodal hemoglobins may, in fact, be components which occur in the embryo and which are retained into the adult phase, just as Hb A6–A8 are in coho salmon. The hemoglobins of the two species of fish differ in their functional characteristics. The hemoglobins of rainbow trout adults have a lower O_2 affinity and a very much higher Bohr shift than embryonic hemoglobins. Coho salmon adult hemoglobins have a low O_2 affinity and a very small Bohr shift while coho fry have a high O_2 affinity and a very large Bohr shift (Iuchi, 1973; Giles and Randall, 1980).

In masou salmon, *O. masou* and brook trout *Salvelinus fontinalis,* Iuchi *et al.* (1975) observed 9 and 12 anodally migrating hemoglobins, respectively, in yolk sac alevins. The patterns were distinct from those of swimming fry and of 4-month-old fish and some components must be true embryonic hemoglobins. Preliminary studies in this author's laboratory indicate that embryonic hemoglobins of anodal electrophoretic mobility also occur in yolk sac embryos of the Atlantic salmon, *Salmo salar*. Thus, although embryonic hemoglobins have not been systematically investigated in many species, they have been reported in at least one member of each of the major salmonid genera.

VI. Comparison with Mammals

Hemoglobin ontogeny in salmonids can now be summarized as follows. In a number of species, e.g., *S. gairdneri* and *O. kisutch,* but possibly in all, a group

of embryonic hemoglobins exhibiting anodal electrophoretic mobilities and relatively high O_2 affinities, are synthesized in embryos (probably in blood islands of
the yolk sac) and appear in the blood circulation of embryos and posthatch fry.
After hatching, when erythropoiesis has shifted to definitive sites in the fry
proper, the postembryonic hemoglobins appear. In all salmonids these separate
electrophoretically into two groups, one group anodal and the other group predominantly cathodal in mobility. Anodal hemoglobins, some of which may
resemble embryonic hemoglobins in electrophoretic mobility, predominate in
juveniles. The ratio of anodal to cathodal hemoglobins declines progressively
throughout the life cycle in a number of species, but never declines below 50%
even in the largest fish. This size-correlated shift in the Hb A/Hb C ratio reflects
a real increase in the amount of Hb C present. Concomitant with this change in
ratio, changes occur in the number and type of individual components within the
Hb A group and within the Hb C group. These may involve the appearance of
new "adult" components (e.g., Hb C11 and C16 of sockeye salmon), the
disappearance of juvenile components (e.g., Hb A4 of Atlantic salmon), and the
transient appearance and subsequent disappearance of others (e.g., Hb A7 of
Atlantic salmon). The multiple hemoglobins can be explained by the occurrence
of different electromorphs of four types of chains, α^F, β^F which form the anodal
hemoglobins and α^S, β^S which form the cathodal. The ontogenetic variations
reflect changes in the occurrence of different electromorphs within these four
classes.

What analogy or homology exists between these complex salmonid hemoglobins and the well known ontogenetic hemoglobins of mammals? Three kinds
of hemoglobins, embryonic, fetal, and adult, appear sequentially in the blood of
man, the best studied example. Embryonic hemoglobin occurs in nucleated
embryonic erythrocytes formed in the yolk sac and they disappear very early in
development; fetal hemoglobins are found in erythrocytes formed in the fetus
proper and they disappear entirely, or form only minor constituents, in the blood
of normal healthy adults. Adult hemoglobins are the characteristic components
observed in the erythrocytes of adults which are synthesized in erythropoietic
sites in myeloid tissue. Structurally, these different kinds of hemoglobin within
an individual all have the same α chains, but differ in their non-α chains. Kitchen
and Brett (1974) provide a good, brief discussion of hemoglobin ontogeny in a
variety of mammal species.

The hemoglobins observed in the yolk sac embryos of rainbow trout, coho
salmon, and, very probably, Atlantic salmon, masou salmon, and brook trout
appear to represent true embryonic hemoglobins synthesized in yolk-sac tissue
and associated with large, disc-shaped erythrocytes. The postembryonic hemoglobins of salmonids are complex and their homology with the fetal and adult
hemoglobins of higher vertebrates is not easily determined. Because the ratio of
anodal to cathodal hemoglobins is high in juvenile fish and declines steadily with

growth, some authors, in presenting or referring to the studies reviewed above, equate the anodal hemoglobins with the "fetal" hemoglobins of other vertebrates, and the cathodal hemoglobins with adult hemoglobins. This appears to be wrong: the decline in Hb A/Hb C ratio in salmonids may not involve any significant decrease in the actual *amount* of anodal hemoglobin with growth which would be expected of a true fetal hemoglobin. The decreasing Hb A/Hb C ratio does reflect a real increase in the amount of Hb C, the putative adult hemoglobin. But Hb C never exceeds 50% of the total hemoglobin even in the largest fish, a value that seems low for a true adult hemoglobin. Finally, the complex ontogenetic changes, both quantitative and qualitative, which are observed in salmonids occupy an extended period of the life cycle (indeed the whole life cycle in some cases) and appear to be well divorced from any detectable larval–adult change. All that can be safely concluded is that those hemoglobins, both anodal and cathodal, which appear latest and remain in the circulation of large, sexually mature individuals are true adult hemoglobins, whereas those hemoglobins, anodal and cathodal, which are present in young fish and decline in intensity gradually over the life cycle are best regarded as juvenile hemoglobins. The transient hemoglobins seen in growing salmon may have no mammalian equivalent.

VII. Control of Hemoglobin Ontogeny

Embryonic hemoglobins, produced in yolk-sac erythropoietic tissue, soon disappear from the circulation of fry after the yolk-sac has been resorbed and the site of hemoglobin synthesis has shifted to the liver and spleen. The qualitative and quantitative ontogenetic variations in the postembryonic hemoglobins have not been ascribed to any such change in the site of erythropoiesis. The ontogenetic rate appears to be an innate biological property of the species and the factors which control it, and how they operate, are at present unknown. The normal rate of decline in the Hb A/Hb C ratio may, of course, be halted, or even reversed, when the fish is stressed by external injury (Koch *et al.*, 1966). Altered levels of many normal blood and cellular substances are not unusual in injured or otherwise stressed organisms. On the other hand, Giles and Vanstone (1976) showed that artificial environmental regimes in which temperature, salinity, and dissolved oxygen were set at different levels (temperature varied from 1 to 15°C, salinity 10–30%, and dissolved oxygen from 3 to 8 ppm) did not cause modification of the hemoglobin pattern in coho salmon fry and presmolts. However, they did observe that individuals kept for 60 days at temperatures of 15°C, or in salinity of 15%, grew faster than individuals reared normally, but their Hb A/Hb C ratio was retarded: in normal fish of approximately 12 cm in length Hb A constituted 72% of the hemoglobin whereas in treated fish of the same length

HbA constituted 80–85% of the total hemoglobin. They concluded that physical size was not the main determinant of ontogenetic change in coho salmon. This point had been made earlier for Atlantic salmon by Koch *et al.* (1964), who stressed that it was *relative* rather than *absolute* size which was important: in the Namsen Blanken stock—a landlocked race of dwarf Atlantic salmon—the hemoglobin ontogeny went to completion in fish of much smaller absolute size than in non-dwarf sea-going stock. Thus, the changing hemoglobin pattern reflects a physiological aging process rather than just a physiological index of size alone.

Westman (1970) also showed that the hemoglobin pattern of landlocked Atlantic salmon in Lake Saimaa, Finland, developed at a faster rate than that of sea-going salmon. This suggests that the intrinsic ontogenetic rate is a population, or race, characteristic rather than a species characteristic. Because the subunit structures of the various anodal hemoglobins are known, it is possible to calculate from Westman's data the proportions of the $\alpha^{Fs}(R)$ and $\beta^{Ff}(V)$ subunits present in individuals of different length taken from the land-locked and sea-going stocks. The results are illustrated in Fig. 9. As expected, α^{Fs} and β^{Ff} both decline with increasing length of fish. However, while the rate of change of the α^F chains is clearly the same in the landlocked and the sea-going populations, that of the β^F chains is much more rapid in the former population. When these two populations are compared at the subunit level, then, they differ in the rate of switching the β^F chains only.

Another factor, other than relative size and race of origin, which influences the observed ontogenetic rate is the process of sexual maturation. Wilkins (1983) reexamined the data of Wilkins (1968) and divided the fish into two categories. One category was comprised of "grilse," i.e., fish returning to freshwater to spawn having spent only 1 year in sea water. The second category consisted of those individuals which did not return to spawn until more than 1 year had been spent at sea. The hemoglobin pattern of the grilse was observed to be further advanced than that of nongrilse fish of the same length. For the anodal hemoglobins, the proportions of $\alpha^{Fs}(R)$ and $\beta^{Ff}(V)$ were calculated exactly as for Fig. 9, and the results (Wilkins, 1983) are presented in Fig. 10. It is clear that while the ontogenetic switch of the β^F chains occurs at the same rate in the grilse and nongrilse categories, the switch of the α^F chains is much faster in the former.

It is possible that this latter observation, made on wild fish caught at sea and in freshwater, could be explained by differences in the location or season of capture, nutritional status of individuals, or some other extraneous factor. For this reason, Wilkins (1983) analyzed a single cohort of salmon which had been hatched and reared as a group throughout the whole life cycle. These were studied when all individuals were between 40 and 60 cm long, having spent one winter together in a sea-cage.

The proportions of α^{Fs} and β^{Ff} declined with increasing length of fish over the relatively short length range studied, but the regression was statistically significant only in those fish classified on external features as "grilse." These exhibited rates

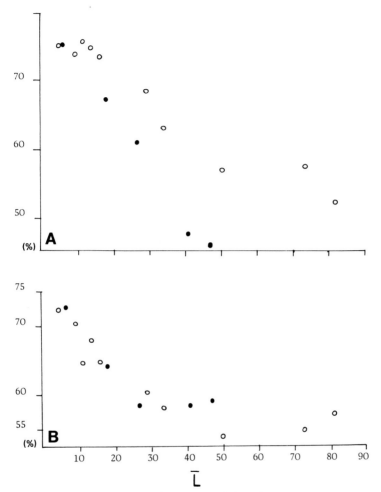

FIG. 9.   Changes in the proportions of the subunit chains of the anodal hemoglobins of land-locked (closed symbols) and sea-going (open symbols) salmon from Finland. (A) β^{Ff} as a percentage of ($\beta^{Ff} + \beta^{Fs}$); (B) Lower graph: α^{Fs} as a percentage of ($\alpha^{Ff} + \alpha^{Fs}$). (Original based on data in Westman, 1970.)

of change greater than those of "nongrilse" of similar size in the same cohort. Within a single cohort, then, those individuals which exhibit the "grilse" phenotype have a higher rate of hemoglobin ontogeny, independently of length and chronological age, than those which do not. It seems that the onset of normal sexual maturation and the development of the hemoglobin pattern are both related to physiological age rather than to chronological age or physical size.

Finally, two important studies of the influence of diet on hemoglobin ontogeny

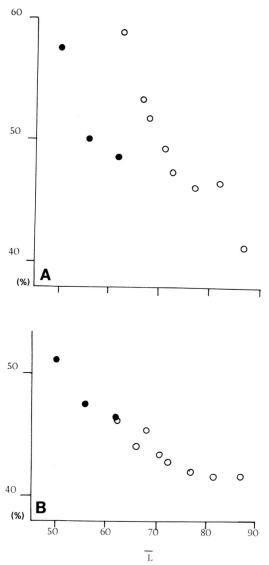

FIG. 10.   Changes in the proportions of the subunit chains of the anodal hemoglobins of grilse (closed symbols) and nongrilse (open symbols). (A) α^{Fs} as a percentage of ($\alpha^{Ff} + \alpha^{Fs}$); (B) β^{Ff} as a percentage of ($\beta^{Ff} + \beta^{Fs}$). (Redrawn with permission from Wilkins, 1983.)

have been reported. Koch *et al.* (1964) showed that young Atlantic salmon which received 20% beef thyroid tissue in their diet displayed hemoglobin patterns well in advance of that expected for their size. This acceleration of the ontogeny was independent of size and sexual maturation. Bergstrom and Koch (1969) reported that when hatchery-maintained young salmon were fed on dry, pelleted food there was a shift in the hemoglobin pattern to a more advanced stage which was not observed in fish maintained on a wet food diet. This result suggested to the authors that young fish fed on dry, pelleted food exhibited accelerated physiological aging compared with siblings fed wet food.

VIII. Ontogeny, Genetics, and Evolution

On the reasonable assumption that each electrophoretically distinct subunit polypeptide chain is encoded at a separate structural gene locus, there are at least 8 distinct structural genes for globin detectable in *O. nerka,* 7 in *S. gairdneri,* 6 in *S. clarkii,* and 8 in *Salmo salar.* If the multiple haemoglobins of the other salmonid species listed in Table I have subunit structures similar to these, then a minimum of 6–8 structural genes for postembryonic hemoglobins must be characteristic of the salmonids. Even more globin genes must be postulated if embryonic hemoglobin is considered.

The polypeptide chains of the postembryonic hemoglobins are of four distinct classes viz. α^F, β^F, α^S, and β^S classes. Summarizing the data in Table II, it is clear that at least two electrophoretically distinct, nonallelic polypeptides occur within each polypeptide class, e.g., α^{Ff}, α^{Fs}. The hemoglobins of mammals are composed of polypeptides of two classes, the α and the non-α classes. The non-α class arose by gene duplication and mutation from the α precursor, and the various γ, δ, β, chains of the non-α class arose by later duplications and mutation as illustrated earlier (Fig. 1). The α, β, γ, and δ genes have each undergone subsequent duplication in a number of species (Kitchen, 1974; Kitchen and Brett, 1974). The occurrence of these evolutionary events can be inferred from studies of amino acid sequence of the globins and the degree of their structural similarity. Chains which resemble each other closely have arisen through an evolutionary recent duplication event; those which show less structural resemblance arose through earlier duplication. Human β- and δ-globins, for example, differ from each other by 10 amino acid substitutions and so are considered to be derived from an older duplication than the two human γ chains, $^G\gamma$ and $^A\gamma$ which differ by only a single amino acid. The α chain duplications of man and other mammals exhibit no detectable electrophoretic differences and may be the most recent duplications (Vogel and Motulsky, 1979).

This structural gene duplication–sequential mutation model provides a clear and logical hypothesis for the occurrence of multiple hemoglobins in many animal species (Kitchen, 1974). With the constraints discussed below the salmonid data fit the model well. Although no amino acid sequence data or DNA sequence data are yet available, the degree of similarity of the various salmonid globin chains can be reasonably inferred. The various chains within a class, for instance the β^F class of *S. salar,* migrate closely together in acid-urea gels and they occur interchangeably when forming tetramers involving the class, e.g., $\alpha_2^{Ff}\beta^{Ff}\beta^{Fs}$. They are, therefore, inferred to be more closely related than, for example, β^{Ff} and β^{Sf} which cannot occur together in the same tetramer.

In this way, two possible scenarios for the evolution of salmonid hemoglobins can be constructed, as in Fig. 11. In both schemes, the intraclass genes, e.g., α^{Ff} and α^{Fs} are assumed to have arisen by gene duplication and mutation from the nominative gene of that class. The β^{Fm} gene of *O. nerka* could have arisen by duplication of either the β^{Bf} or the β^{Fs} gene; in both schemes it is presented as a β^{Ff} duplication simply for consistency. Both schemes postulate an evolutionary phase in which there existed only one gene for each class of globins. How these may have arisen differs in the two schemes. In A they are presented as originating by duplication and mutation from putative ancestral genes labeled pro-F and pro-S which encoded anodal and cathodal products respectively. In B they are presented as duplications from ancestral pro-α and pro-β genes; the electrophoretic mobility of whose products is unknown. In both schemes the "pro-" genes are presented as arising by duplication of a single ancestral gene at the very beginning of vertebrate evolution. B resembles closely the model of higher vertebrate hemoglobin evolution: the pro-β gene arises as a duplication of an $\alpha\beta$ ancestor and both lines experience later duplications to give the present-day gene array. This scheme cannot be easily discounted. Nevertheless, A has obvious merits in explaining the observed hemoglobin phenotypes. Within the Hb A group, and within the Hb C group, tetrameric hemoglobins involve quaternary interactions of α and β subunits, but cross-interactions are never observed between the two groups. For example, in forming a tetramer with α_2^{Ff}, two β^F polypeptides are required. They cannot be replaced or substituted by β^S polypeptides in any combination. Because the α and β chains within the F(Hb A) or S(Hb C) groups can interact together, but they are not capable of cross-interaction, then the α^F and β^F classes are regarded as functionally more closely related to each other than either is to the α^S or β^S class. A recognizes this functional relatedness by deriving the α^F and β^F genes from a single common ancestral gene (pro-F) and the α^S and β^S genes from a different ancestral gene (pro-S). In B, genes whose products are capable of functional interaction, e.g., α^F, β^F are presented as less closely related by descent than genes not capable of functional interaction, e.g., α^F, α^S. Because A better correlates functional relatedness with relationship by descent, it is preferred by this author.

The final duplication leading to nine detectable structural loci is not electrophoretically evident in all the salmonid species; in some instances the products

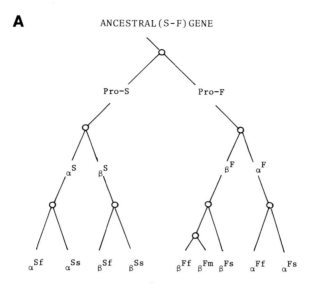

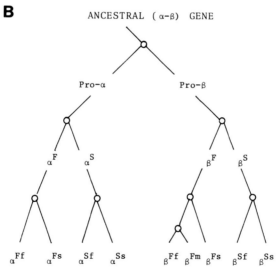

FIG. 11. Hypothetical schemes for the evolution of hemoglobin structural genes in salmonids. (Original). (A) Duplication of Pro-S and Pro-F genes. (B) Duplication of Pro-α and Pro-β genes. ○, Gene duplication.

of the recently duplicated loci may still be structurally identical. The biochemical evidence for extensive gene duplication at other loci in salmonids is impressive and the group is considered to have undergone a tetrapoidization event in the course of its evolution (Ohno *et al.,* 1968). The duplication leading to separate α- and non-α-globin genes occurred in the ancestral vertebrates about 380×10^6 yr B.P. (Vogel and Motulsky, 1979). The salmonids evolved from their putative ancestors with the Ganoids about 160×10^6 yr B.P. (Tchernavin, 1939). Thus the tetrapoidization event in the evolution of the ancestor of modern salmonids occurred after the ancestral α–β duplication of vertebrates. In A, then, the tetraploidization is represented by the third (latest) level of duplication; in B, tetraploidization may be represented by either the second or third level of duplication.

Once duplicated genes had arisen and diverged, the possibility of the regulation of structural gene expression arose. Evolution of gene regulatory mechanisms is an important aspect of organismic evolution (Markert *et al.,* 1975), although the process is still only poorly understood in eukaryotes. In humans, the change from fetal to adult hemoglobin, involving the repression of γ chain production and the stimulation of β chain production, is controlled by nucleotide sequences in, and close to, the γδβ structural gene cluster (Flavell and Grosveld, 1983). In particular, a nucleotide sequence situated between the ^{A}γ and δ structural genes on the 5′ side of the β-gene is important. Such nucleotide sequences form part of the regulatory mechanism or "regulatory gene" which controls the sequential expression of the γ, β structural genes.

Among the salmonids, the observed hemoglobin ontogeny is consistent with control by regulatory genes acting independently on each of the four globin classes ($α^F$, $β^F$, $α^S$, $β^S$). Regulation is not necessarily observed, nor is it expected, in all species or in all globin classes within a species (see Table II). It cannot occur, for example, where intraclass gene duplication or divergence has not occurred, as in the $α^F$ class in sockeye salmon (Fig. 3). The sequence of events in the evolution of regulated genes consists of structural gene amplification followed by structural divergence followed by the evolution of regulatory control (Markert *et al.,* 1975). All phases of this evolutionary sequence in the $α^S$-globin class are represented among the species listed in Table II. The Atlantic salmon *Salmo salar,* having two distinct structural genes within all four globin classes and regulatory control on each, exhibits the most highly evolved hemoglobin genome. In all populations of Atlantic salmon throughout the whole of its range the same hemoglobin patterns are observed. Only the rates at which the subunits are regulated differ between populations. The rate also differs between grilse and 2 sea-winter salmon within the population of a single river. Is this an example, within a single species, of evolution by control gene divergence? Is the extended ontogeny of salmonid hemoglobin homologous with, and an evolutionary forerunner of, the fetal–adult shift of higher vertebrates? At what level, and in what manner, does dietary thyroid act to alter the intrinsic ontogenetic rate? These and many other questions remain unanswered for the present.

ACKNOWLEDGMENTS

I am grateful to the publishers of *Journal of the Fisheries Research Board of Canada* for permission to reproduce Fig. 2, the publishers of *Comparative Biochemistry and Physiology* for Figs. 3, 7, and 8, the publishers of *Biochimica et Biophysics Acta* for Figs. 5 and 6, and the publishers of *Irish Fisheries Investigations* for Fig. 10.

REFERENCES

Axel, R., Maniatis, T., and Fox, C. L., eds. (1979). "Eucaryotic Gene Regulation." Academic Press, New York.

Bergström, E., and Koch, H. J. A. (1969). *Swed. Salmon Res. Inst. Rep., LF1 Medd.* **9**, 1–7.

Buhler, D. R., and Shanks, W. E. (1959). *Science* **129**, 899–900.

Diesseroth, A., Nienhuis, A. W., Turner, P., Velez, R., Anderson, W. F., Ruddle, F., Lawrence, J., Creagan, R., and Kucherlapati, R. (1977). *Cell* **12**, 205–218.

Diesseroth, A., Nienhuis, A., Lawrence, J., Giles, R., Turner, P., and Ruddle, F. H. (1978). *Proc. Natl. Acad. Sci. U.S.A.* **75**, 1457–1460.

Eguchi, H., Hashimoto, K., and Matsuura, F. (1960). *Bull. Jpn. Soc. Sci. Fish.* **26**, 810–813.

Flavell, R. A., and Grosveld, F. G. (1983). "Eucaryotic Genes: Activity and Regulation." Butterworths, London, in press.

Flavell, R. A., Bernards, R., Grosveld, G. C., Hoeijmakers-Van Dommelen, H. A. M., Kooter, J. M., De Boer, E., and Little, P. F. R. (1979). *In* "Eucaryotic Gene Regulation" (R. Axel, T. Maniatis, and C. F. Fox, eds.), pp. 335–354. Academic Press, New York.

Giles, M. A., and Randall, D. J. (1980). *Comp. Biochem. Physiol.* **65A**, 265–272.

Giles, M. A., and Vanstone, W. E. (1976). *J. Fish Res. Bd. Can.* **33**, 1144–1149.

Hashimoto, K., and Matsuura, F. (1959a). *Bull. Jpn. Soc. Sci. Fish.* **24**, 719–723.

Hashimoto, K., and Matsuura, F. (1959b). *Nature (London)* **184**, 1418.

Hashimoto, K., and Matsuura, F. (1959c). *Bull. Jpn. Soc. Sci. Fish.* **24**, 724–729.

Hashimoto, K., and Matsuura, F. (1959d). *Bull. Jpn. Soc. Sci. Fish.* **25**, 465–469.

Hashimoto, K., and Matsuura, F. (1960a). *Bull. Jpn. Soc. Sci. Fish.* **26**, 354–360.

Hashimoto, K., and Matsuura, F. (1960b). *Bull. Jpn. Soc. Sci. Fish.* **26**, 931–937.

Hashimoto, K., and Matsuura, F. (1962). *Bull. Jpn. Soc. Sci. Fish.* **28**, 914–919.

Hashimoto, K., Yamaguchi, Y., and Matsuura, F. (1960). *Bull. Jpn. Soc. Sci. Fish.* **26**, 827–834.

Iuchi, I. (1973). *Comp. Biochem. Physiol.* **44B**, 1087–1101.

Iuchi, I., Suzuki, R., and Yamagami, K. (1975). *J. Exp. Zool.* **192**, 57–64.

Kitchen, H. (1974). *Ann. N.Y. Acad. Sci.* **241**, 12–24.

Kitchen, H., and Boyer, S. H., eds. (1974). *Ann. N.Y. Acad. Sci.* **241**, 1–737.

Kitchen, H., and Brett, I. (1974). *Ann. N.Y. Acad. Sci.* **241**, 653–671.

Koch, H. J. A. (1982). *Aquaculture* **28**, 231–240.

Koch, H. J. A., Bergström, E., and Evans, J. C. (1964). *Meded. K. Vlaam. Acad. Wet. Lett. Schone Kunsten Belg. Kl. Wet.* **26**, 1–32.

Koch, H. J. A., Bergström, E., and Evans, J. C. (1966). *Meded, K. Vlaam. Acad. Wet. Lett. Schone Kunsten Belg. Kl. Wet.* **28**, 1–20.

Koch, H. J. A., Wilkins, N. P., Bergström, E., and Evans, J. C. (1967). *Meded. K. Vlaam. Acad. Wet. Lett. Schone Kunsten Belg. Kl. Wet.* **29,** 1–16.

Maniatis, T., Butler, E. T., Fritsch, E. F., Hardison, R. C., Lacy, E., Lawn, R. M., Parker, R. C., and Shen, C. (1979). *In* "Eucaryotic Gene Regulation" (R. Axel, T. Maniatis, and C. F. Fox, eds.), pp. 317–333. Academic Press, New York.

Markert, C. L., Shaklee, J. B., and Whitt, G. S. (1975). *Science* **189,** 102–114.

Ohno, S., Wolf, U., and Atkin, N. B. (1968). *Hereditas* **59,** 169–187.

Ronald, A. P., and Tsuyuki, H. (1971). *Comp. Biochem. Physiol.* **39B,** 195–202.

Schumann, G. O. (1959). *Rep. Inst. Freshwater Res.,* Drottningholm, **40,** 176–197.

Stamatoyannopoulos, G., and Nienhuis, A. W., eds. (1979). "Cellular Molecular Regulation of Haemoglobin Switching." Grune & Stratton, New York.

Tchernavin, V. (1939). *Salmon Trout Mag.* **95,** 120–140.

Tsuyuki, H., and Gadd, R. E. A. (1963). *Biochim. Biophys. Acta* **71,** 219–221.

Tsuyuki, H., and Ronald, A. P. (1971). *Comp. Biochem. Physiol.* **39B,** 503–522.

Vanstone, W. E., Roberts, E., and Tsuyuki, H. (1964). *Can. J. Physiol. Pharmacol.* **42,** 697–703.

Vogel, F., and Motulsky, A. G. (1979). "Human Genetics." Springer-Verlag, Berlin and New York.

Westman, K. (1970). *Suom. Tiedeakat. Toim. Ser. A, IV, Biol.* **170,** 1–28.

Wilkins, N. P. (1968). *J. Fish. Res. Bd. Can.* **25,** 2651–2663.

Wilkins, N. P. (1970). *Biochim. Biophys. Acta* **214,** 52–63.

Wilkins, N. P. (1983). *Irish Fish. Invest. Ser. A* No. 23, 67–71.

Yamamoto, M., and Iuchi, I. (1975). *J. Exp. Zool.* **191,** 407–426.

Yamanaka, H., Yamaguchi, K., Hashimoto, K., and Matsuura, F. (1967). *Bull. Jpn. Soc. Sci. Fish.* **33,** 195–203.

Yoshiyasu, K. (1973a). *Bull. Jpn. Soc. Sci. Fish.* **39,** 97–114.

Yoshiyasu, K. (1973b). *Bull. Jpn. Soc. Sci. Fish.* **39,** 449–459.

Yoshiyasu, K., and Humoto, Y. (1972). *Bull. Jpn. Soc. Sci. Fish.* **38,** 779–788.

Index

A

ABP, *see* Androgen binding protein
Acrocentric associations, 160–162, 167–170
Adenylate cyclase, 141
Adluminal compartment, 181
Androgen binding protein, 129, 133, 134–138, 143–145

B

Barley, CMS in, 256–257
Basal body, *see* Polar organelles
Basal compartment, 181
Blood–testis barrier, 128–130, 183

C

Carcinogenesis, chromatin transformation during, 44–45
Carrot, CMS in, 254–255
Cell cycle, chromatin organization and, 42–44
Centriole, *see* Polar organelles
Centromere, 77, 102
  replication, 92–93
Chromatin, *see also* Condensation
  condensed, 25, 26
  in dinoflagellates, 11
  replication, 41–42
Chromatin fibers, 23–25
Chromatin organization, 21–56
  of active genes, 37–38
  carcinogenesis and, 44–45
  cell cycle and, 42–44

genetic factors and, 30
levels of, 22–29
mammalian differentiation and, 35–37
models of gene regulation, 45–49
nucleotypic factors and, 29–30
plant differentiation and, 31–35
in plants, 29–35
Chromocentric nucleus, 27
Chromomeric nucleus, 27
Chromonematic nucleus, 27
Chromosomal disorders, 162–166
Chromosomes, dinoflagellate, 5–19
  chemical composition, 8–10
  helical compaction, 14
  number of, 7–8
  replication and division, 17
  structural organization, 11–16
  three-dimensional structure, 11–15
Chromosome, human, nucleolar organizer regions, 151–176
Chromosomes, plant
  classification of, 1–3
  evolution of, 1–2
  functional, isolation of, 120–121
    manipulation of, 121–124
  linearization, 108–109
  metaphase, 57–76
  nonhistone proteins, 59–60
  nuclear location, 107–108
  protein-depleted, structure of, 60–66
  rearrangement of, during differentiation, 115–118
Chromosomes, vertebrate, metaphase, 69–74
Close junction, 194–195

Contents of Recent Volumes